Power Button

Power Button

A History of Pleasure, Panic, and the Politics of Pushing

Rachel Plotnick

The MIT Press
Cambridge, Massachusetts
London, England

This book was set in ITC Stone Sans Std and ITC Stone Serif Std by Toppan Best-set Premedia Limited.

Library of Congress Cataloging-in-Publication Data

Names: Plotnick, Rachel, author.
Title: Power button : a history of pleasure, panic, and the politics of pushing / Rachel Plotnick.
Description: Cambridge, MA : The MIT Press, [2018] | Includes bibliographical references and index.
Identifiers: LCCN 2017055846 | ISBN 9780262038232 (hardcover : alk. paper), 9780262551953 (paperback)
Subjects: LCSH: Remote control--Psychological aspects--History. | Electric switchgear--Psychological aspects--History. | Social psychology. | Industrial Revolution. | Object (Philosophy)
Classification: LCC TJ213.5 .P55 2018 | DDC 621.3815/37--dc23 LC record available at https://lccn.loc.gov/2017055846

Contents

Acknowledgments

I am humbled by and grateful for many individuals' contributions and generosities that made this book possible.

The project found its voice in the archives—in the pages and images of the past. Because no folder or box labeled "Button" exists in any archive, however, I relied on the continuous support provided by archivists, librarians, and other research staff. In particular, I am indebted to the National Museum of American History Archives Center at the Smithsonian Institution in Washington, DC, and express my thanks to archivists Alison Oswald and Kay Peterson. I also received a research grant from NMAH's Lemelson Center for the Study of Invention and Innovation, and I appreciate the support I received from the Center and from Eric Hintz. A second important site of research, the Bakken Museum and Library in Minneapolis, MN, offered a wealth of resources, and I am grateful for the enthusiastic assistance that librarian Elizabeth Ihrig gave me during my time there. In addition, I interacted with helpful staff at the Connecticut Historical Society in Hartford, CT; Huntington Library in San Marino, CA; and Winterthur Library in Winterthur, DE.

Before I began hunting and pecking through archives, this project arose out of curiosity after a seminar I took while completing my PhD in the Media, Technology, and Society (MTS) program at Northwestern University's School of Communication. As it took further shape in the form of a dissertation, I benefited immensely from the intellectual community and mentors there who embraced the concept. Jennifer Light, my advisor, provided essential feedback at every stage of the process, not only helping me to work through my thoughts and improve my writing but also to encourage me personally and professionally. I am tremendously appreciative of her continued guidance and support well beyond the boundaries of our time at Northwestern. This book surely would not exist without her, and my gratitude is simply immeasurable. Pablo Boczkowski consistently pushed me forward by exposing me to generative bodies of literature, asking me difficult and important questions, and advising on myriad matters of professionalization. I continue to admire and have been enriched by his work ethic, collegiality, and support of his colleagues and students. Lynn Spigel invested significantly in this project, and her confidence spurred me forward at every stage. I am particularly appreciative to Lynn for suggesting early on that I present my work at Northwestern's Medium to Medium symposium, which provided fertile ground for testing out my ideas.

In addition to these exceptional mentors, I wish to express my sincere gratitude to other faculty members at Northwestern, especially Ken Alder, James Ettema, Eszter Hargittai, and Janice Radway, all of whom enriched my understanding of major scholarship and sharpened my critical thinking. At the same time, I benefited from a wonderful group of fellow graduate students. My heartfelt thanks go to Will Barley, Alan Clark, Lindsay Fullerton, Katie Day Good, Yuli Patrick Hsieh, Nicole Joseph, Aditi Raghavan, Lauren Scissors, Ignacio Siles, Kristin Yates Thomas,

Jeffrey Treem, Brooke Foucault Welles, Angela Xiao Wu, and many others.

As I transitioned to a faculty position at the University of North Carolina at Charlotte, I was fortunate enough to join an exceptional department of colleagues who supported me from the start. Kind and encouraging words as well as feedback in various forms, especially from Jon Crane, Dan Grano, and Min Jiang, have motivated me throughout the writing and editing process. Jamie Bochantin, Cris Davis, Margaret Quinlan, Robin Rothberg, Cliff Scott, and Ashli Stokes have especially made the department a welcoming and collegial intellectual home. I also appreciate continued guidance from Jason Black, Richard Lee-man, and Shawn Long, each of whom have invested not only in this book but also in my success as a scholar.

Outside of the department, this project found its "legs" at many interdisciplinary annual meetings of such organizations as the International Communication Association, National Communication Association, Society for Cinema and Media Studies, Society for the History of Technology, and Society for the Social Studies of Science. In these forums and others, I have had the good fortune of learning from, listening to, and collaborating with faculty members from across the country. I extend my sincere thanks to these generous individuals, including Ruth Schwartz Cowan, Barbara Friedman, Bernard Geoghegan, Robert MacDougall, Benjamin Peters, Jeff Pooley, Rick Popp, Fred Turner, and numerous others. Likewise, journal editors Suzanne Moon and William Aspray invested significant time in providing feedback on portions of this work, and I appreciate the many ways that they helped me to develop my ideas. In addition, special thanks go to David Parisi, who has challenged me with stimulating conversations and provided critical support for my work over the years as we've worked together to push "haptic media studies" forward.

I am incredibly grateful to my editor Katie Helke, who has served as a patient, thoughtful, and dedicated guide. Her feedback, along with that of three anonymous reviewers, has enabled me to see the topic with fresh eyes and to improve my writing in ways that I could not have imagined at the outset. I also thank the numerous staff at the MIT Press who have shepherded this book through every stage of the publishing process.

Finally, it is difficult to fully express my appreciation to friends and family who have graciously encouraged me personally and professionally. My dear friend Lissy Skolnick has lent an ear—both short and long-distance—over the years with incredible kindness and generosity of spirit. Similarly, my love and thanks go to Sue Holland, Phil Holland, and Mary Franklin, who have served as champions of this book (and everything I do).

My brother, Daniel Plotnick, has always good-naturedly listened and offered his warmest support, talking me through difficult times as only a brother can. He and I have achieved what we have because of our parents' love. A thank you feels insufficient for my mom, Robin Plotnick, who has walked every step with me and is a source of unending strength. She believed that this book would come to be even when I could not, and she has contributed every bit of her time and help to make it possible. As I write this, I also think of my dad, who always encouraged me to "write, write, write." His absence is felt every day, and I carry him in the closest part of my heart.

I can trace the genealogy of this book to the timeline of two pregnancies, to both of my children's births, and to each of their early years. Shane and Maya have given me the gift of being their mother, and I am awestruck to have such good fortune. This luck is only multiplied by the fact that I share them with my partner and best friend, Stuart Holland. He has made my dreams his own and sustained me with his humor and unwavering support, and I dedicate this book with love and gratitude to him.

Earlier versions of parts of this book appeared in the following publications:

Rachel Plotnick. "At the Interface: The Case of the Electric Push Button, 1880–1923." *Technology and Culture* 53, no. 4 (2012): 815–845. © 2012 The Society for the History of Technology. Reprinted with permission by Johns Hopkins University Press.

Rachel Plotnick. "Force, Flatness and Touch without Feeling: Thinking Historically about Haptics and Buttons." *New Media & Society* 19, no. 10 (2017): 1632–1652. © 2017 SAGE Publications. Reprinted by permission of SAGE Publications. https://doi.org/10.1177/1461444817717510.

Rachel Plotnick. "Panic Button: Thinking Historically about Danger, Interfaces, and Control-at-a-Distance." In *Communication and Control: Tools, Systems, and New Dimensions*, edited by Robert MacDougall, 45–57. London: Lexington Books, 2015. Reprinted by permission of Rowman & Littlefield.

Rachel Plotnick. "What Happens When You Push This? Toward a History of the Not-So-Easy Button." *Information & Culture: A Journal of History* 50, no. 3 (2015): 315–338. © 2015 University of Texas Press. All rights reserved.

Rachel Plotnick. "Touch of a Button: Long-Distance Transmission, Communication, and Control at World's Fairs." *Critical Studies in Media Communication* 30, no. 1 (2013): 52–68. © National Communication Association. Reprinted by permission of Taylor & Francis Ltd. (http://www.tandfonline.com) on behalf of National Communication Association.

Introduction

So far as the great engines were concerned, it made no difference whose finger "pressed the button," but the people began to realize that ... it makes a vast difference who "pressed the button."
—Geo. L. Cooper, 1893[1]

Today, from the "big red button" and Staples' "Easy Button" to Facebook's "Like" button, push buttons loom large in our cultural imaginary, with a vision of button pushing that always looks the same: push a button and something magical begins. A sound erupts that seems never to have existed before. A bomb explodes. A vote registers. A machine animates, whirling and processing. A trivial touch of a single finger sets these forces in motion. The user is all powerful, sending the signal that turns on a television, a mobile phone, a microwave. She makes everything go. Whether or not she understands how the machine works, she determines the fate of the universe. For more than a century, this supposed asymmetry between a minor touch of the hand and its swift effects has cast push-button interactions as both desirable and dangerous. The symbolic, seductive worlds of push buttons make great fodder for novels, films, and advertisements, in which exceedingly simple buttons never fail and any hand can push them.

We are surrounded by buttons—new products that aim to attract users who desire an effortless technical experience appear daily. Marketing efforts rely upon buttons to sell not only a particular kind of interactivity, but also a way of life, from Dell's "Productivity at the touch of a button" slogan advertising its Latitude ON | FLASH module, which allows busy workers to quickly awaken a sleeping computer to perform basic functions such as checking email or performing a web search, to BMW's "comfort access system" promising that "so much at your fingertips" is made possible by flipping a single switch. Similarly, PayPal offers a "One Touch" system to make online payments, Amazon Dash buttons promise that one can "Just press the button to get your essentials," and Uber enthuses that its customers can "Tap a button. Get a ride." Sometimes these "buttons" are not only functional but also symbolic, as in the case of the US government's "Blue Button," an image displayed on patient portals and secure websites to help individuals gain access to and securely share their medical information. Meanwhile, a click or tap of a button on nearly any social media site allows for the "buttonization" of emotions—one can express a feeling much as they might choose a snack from a vending machine.[2] This small sampling of the prevalence of buttons, which doesn't even begin to account for the button pushing required to call an elevator or take a selfie, tells us a great deal about the current moment and also demonstrates the remarkable durability of buttons—both physical and imagined—over time, despite persistent efforts to replace them. Why are buttons so ubiquitous? Why do people love and loathe them? How do they work? In what ways do they interact with societal understandings of safety, pleasure, danger, or politics? How and when does "pushing" act as a form of force or coercion when a finger on the button sets someone or something into motion? This book aims to answer these questions by detailing how buttons get made, distributed, used, rejected, and refashioned throughout history, drawing on the past as a window into understanding the present. Taking a historical approach, it argues that now is the time to counterbalance a predominant myth of simplicity with another perspective that treats neither hands nor buttons nor the relationship between the two as neutral or natural. In actuality, unpredictable buttons break, fail, and require repair. They confuse and frustrate. People actively work to prevent button pushing and limit certain kinds of touches by creating technical safeguards against unruliness and so-called misuse. Those who push buttons often do so to maintain privilege, exercise power, and enforce hierarchies.

In the chapters that follow, I argue that hands and machines began to work together in new ways that challenged previous definitions of "manual" labor, "craft," and touch. To date, histories of industrialization have described the shift from manual labor to machines as a totalizing one, noting that machines replaced hands, or they have focused on hands as the tools of hard physical labor.[3] Very few works examine the complexities of hand–machine relationships, and those that do discuss touch broadly as a sensation rather than considering the hand or the finger; they are primarily concerned with philosophical and theoretical inquiries about touch rather than historicizing hand practices.[4] Similarly, cultural histories of electrification, communication, and labor have largely overlooked a society enamored with the "digital"—the finger—as a source of tactile input for machines, turning instead to the sights and sounds of the era. Beyond buttons specifically, a wealth of "technologies of the hand" proliferated that included typewriters, telegraphs, and fingerprinting. These technologies featured interventions to make hands increasingly efficient, to mechanize their activities, and to create an established science around hands' capabilities. Just as modes of listening, hearing, or seeing may have changed, creating new kinds of "soundscapes" and visual spectacles, so did acts of touching—and the very definition and purpose of hands—undergo a process of reimagining that deserves attention.⁵

Power Button suggests that to operate a machine with one's hands necessitated not only inventing a new form of control (which took shape as a button), but it also gave rise to a new kind of controller who used fingers to delegate tasks to other humans and machines in ways that were designed to be simplistic, ergonomic, and at a distance. I propose that this shift in thinking about what hands could do with machines produced an identity for a certain group of users: people who communicated and controlled with minimal effort from their fingertips. This management style, which rose to prominence in the late 1800s, aimed to make people and electrical forces controllable, responsive, and instantly present—or strategically invisible—at the push of a button. I refer to the bundle of technologies, hand practices, environments, and protocols that coalesced around button pushing as *digital command*—an idealized (and often resisted) way of thinking about service, labor, and human–machine relationships.

Electricians, architects, manufacturers, advertisers, and scientific management enthusiasts most often espoused the benefits of digital command, which included a number of commonly referenced components. First, they circulated ideas about hand force or pressure, favoring "mere touch" to groping and straining so that any hand, from a child's to an invalid's, could push a button. Second, they articulated the need for architectural, transportation, and entertainment spaces where a controller would always have buttons within reach, thereby making any space a site for control while putting undesirable elements at a distance. Third, they suggested push-button designs that emphasized concealment, offering an unassuming and discreet method of control and drawing as little attention as possible to the mechanism itself so that push-button effects would appear magically. Last, they proposed foolproof and easy-to-use push-button products oriented toward the nontechnical user. It is important to note that actual activities with push buttons often fell short of, violated, or contested many of these elements. This tension between the promise of digital command and its implementation serves as a central through-line for the chapters that follow. Yet, despite such incongruities, this investment in digital command set the stage for conditions that exist in the present day.

To this end, the book uses the term "digital" intentionally to capture the doubly digital nature of button pushing as an act of the finger (a "digit") and as binary in its activity (buttons worked using a binary logic of two choices such as the "make" and "break" or ON/OFF of electrical current).[6] In addition, today it is impossible to separate the concept of "digital" from computers, a reference to the binary code of 1s and 0s that makes computers function. The tenets of digital command served as a necessary precursor to how we think of the contemporary web-enabled computer user. In the early imaginings of digital command, we witnessed the first ideas of who computer users would be, how they would interact manually (digitally) with computers, and how the ethos of the internet as a whole would grow tied to expectations of consumption, gratification, and access to information at a touch. At the same time, the choice of "command" comes from one of its primary definitions: "an order authoritatively made and remaining in force."[7] This definition captures the fact that button pushers applied force, no matter how gentle, to stir machines and people into action, and they did so by giving an order: stop or start, up or down, on or off, come or go. Despite the guise of politeness, harmony, magic, or ease in button pushing, this essential dynamic between the person pushing and the person or mechanism "pushed" into motion reflects the forceful and commanding nature of the act.

Crafting this history of buttons and the people who pushed them requires first defining what counts in the category of "button," making inevitably complex choices about the boundaries of inquiry. The word "button," from the French *bouton*, originally referred to "a pimple, any small projection" or "to push, thrust forward" beginning in the fourteenth century.[8] The first use of buttons as clothing fasteners dates to a similar time period in Germany; being both fashionable and collectible, they ultimately earned so much attention that, in the early 1800s, Charles Dickens wrote an extended account on their aesthetics and production, deeming them "truly a wonderful and beautiful apparatus."[9] Around this time, inventors also began to describe buttons as objects one could push with a finger. An early example appears in descriptions of Nicolas Rieussec's chronograph (1821), or stopwatch. "When the observer wishes to mark the precise instant when any phenomenon takes place, he presses a button," one journal recounted.[10] Using this "computor," the button "put the machinery in motion" and, likewise, stopped it.[11] Similarly, in some contexts telegraph operators would call the bulbous, rounded end of telegraph keys "buttons," as one would press them to start or stop electric current when sending a message.[12] Another key-based technology, the early typewriter, also featured "finger keys that have at their outer ends buttons or knobs."[13] Although we are primarily concerned here with the act of pushing, other buttons in the vein of dials or knobs required turning but were still referred to as buttons because they protruded from a given surface.[14]

Indeed, tremendous variation existed in the ways that people named buttons and thought about how to use them. Therefore, this book considers buttons to be any of those mechanisms that historical actors called "buttons"; however, it focuses primarily on those buttons that required a single push to create an effect rather than those that required multiple pushes strung together to achieve meaning (as in typewriter keys). What matters most is *why* people called certain devices buttons, what those buttons could do, and how they achieved social and technical relevance in everyday life for those that encountered them. To explore how push buttons—and the fingers that pushed them—came together is to engage with the "messiness" of historical work. How to account for a technology so pervasive yet mundane and often invisible? One historical account might follow musical instruments, beginning at keys of sixteenth-century spinet pianos and transitioning into valves of an early nineteenth-century trumpet.[15] Yet another might emerge from Cold War politics and the mythical "big red button" of nuclear war, following buttons as instruments of power, politics, and social panic.[16] One could catalog the many buttons used for games and play at different moments in time from pinball machines to video games, or identify the pushes at work in various modes of transportation from elevators to trains to automobiles.

It is clear that no single narrative thread could account for all buttons at all times, nor would it tell us much about the underlying processes that have made buttons prominent, meaningful, or pervasive. As a result, *Power Button* borrows from recent traditions that have emphasized "think[ing] historically but avoid[ing] the idea that there are such things as simple origins."[17] It charts a period from approximately 1880 to 1925 to examine the concerns of a first-generation push-button society. These years, characterized by industrialization and electrification, featured movements that often championed push buttons for their roles in summoning whomever or whatever the pusher desired, thereby (theoretically) making tasks increasingly efficient, safe, and comfortable. Whereas few would have engaged with the concept of pushing a button in the mid-nineteenth century, by the 1930s these mechanisms for communication and control had earned a reputation for being ubiquitous and even ordinary. Society became transfixed with the idea that certain people should work and play with their fingers at a comfortable distance from those they commanded. The legacy of this button "mania," centered on digital command, continues to dominate the present moment.

In the late nineteenth and early twentieth centuries, fingers fixed upon push buttons to make doorbells ring, call servants and elevators, turn lights on and off, explode dynamite at a safe distance, and alert police to domestic burglaries. They made toys appear to dance without wires, sent people to their death via electric chair, caused fountains to flow, and operated automobile horns. Someone might encounter a push button in a streetcar, at the front door to a home, at an office desk, in an elevator, or while staying in a hotel. In these circumstances and many others, push buttons functioned as control and communication mechanisms between users and nascent electrical technologies and between human bodies and their surroundings. Electricians and manufacturers imagined that push buttons would remove complexity from technical experiences, hiding away wires, plugs, electricians, batteries, generators, electrical supply companies, inventors, buildings, laborers, and politics behind a nonthreatening button.

Early practices with push buttons most commonly involved calling servants or employees to do the pusher's bidding. In this regard, button pushing acted as a form of summons or demand to conjure one's wishes into being. Button pushers often confronted the complexities of making individuals appear and disappear at a whim. This practice began to change, however, as electric companies endeavored to sell the benefits of electrification, taking great pains to emphasize electricity's invisibility so as to make it appear tame and pliable to the will of human beings. It is unsurprising, then, that the most common metaphor for electricity involved referring to it as an "unseen servant" that responded instantaneously by virtue of a button.[18] Yet despite the promise of "magical" buttons that could conjure anything one desired with a push, fingers, buttons, and a complicated set of technologies and services needed to align to permit the experience of effortless button pushing. Indeed, magical effects could only result when buttons could conceal the "inside" of the machine.[19] Society constructed buttons in complicated ways at a historical moment characterized by tremendous cultural upheaval, reconfigurations of human–machine relationships, and "distinctive new modes of thinking about and experiencing time and space."[20]

It is important to note, however, that most Americans did not have access to push buttons in the period studied in this book. Those with homes wired for electrical services may have encountered them, but only about 8 percent of US households had electricity by 1915.[21] Most users in this early period were by and large part of an urban and affluent population or involved, whether professionally or as skilled "tinkerers," with the electrical community. As a result, industrialization constituted a generative period when the *conditions of possibility* arose to make pushing a button a desirable option, although the push-button design lacked the practicality and widespread applicability that would come in later years. Generally speaking, those who had access to push buttons occupied positions of status: wealthy homeowners, electricians, managers, bankers, and political figures. Conversely, those in positions of servitude such as servants, chauffeurs, entry-level employees, and so on were often made to heed the button's call. Pushing buttons could either empower laypersons by putting new electrical technologies within reach of those groups typically excluded from machine culture (such as children and the elderly or infirm), or disenfranchise laypersons (especially craftspeople, servants, and low-level employees), by placing restrictions on the kinds of machine interactions available to users. In many instances, society constructed the meaning and use of buttons in contradictory ways that both reinforced and defied expectations of gender, class, and race.

Power Button is arranged in three parts that build on one another, detailing a set of practices, ideas, problems, and technologies around the concepts of digital command. While it follows a roughly chronological trajectory, the structure allows for a loose interplay between historical moments so as to make connections that might otherwise go unnoticed. It takes seriously discourse, art, advertising, and fiction at the same time that it discusses patents, catalogs, and wiring schemes. In this regard, it calls for an approach that values how people have talked and thought about buttons as much as how people have constructed, disseminated, or used buttons. Part I, "You Rang?" broadly explores the relationship between button pushers and those individuals made to heed their call, both in terms of hand movements and changing labor dynamics. The section traces how digital commanders used buttons to ring bells for all manner of signaling in order to make requests for attendants' presence increasingly discreet and "polite." Chapter 1, "Setting the Stage," offers initial historical and cultural context to describe how buttons and button pushing achieved prominence. It proposes that the act of pushing a button offered a solution to a number of problems related to labor and effort, perceived dangers of electricity, and control from a distance. Turning to an examination of early button uses, chapter 2, "Ringing for Service," describes how people routinely used bells in the spaces and practices of everyday life to manage interpersonal relations through one-way communication. Elaborating on these activities, chapter 3, "Servants out of Sight," uncovers how tensions grew between those made to work with their hands (manual laborers) and those who commanded with their fingers (digital button pushers). Whereas ringing a bell offered an ergonomically desirable option, putting the button-pushing hand within close proximity of the push, those being summoned expressed frustration at their presence being "toggled" on and off at the pusher's whim.

Part II, "Automagically," shifts to thinking about buttons as tools to spur machines and electrical forces into action with little human intervention. Chapter 4, "Distant Effects," recounts experiments with sending touches across distance to demonstrate how fingers could delegate their efforts to machines in order to produce spectacular, shocking, and sometimes dangerous effects. Where the electrical industry sold a romanticized notion of this version of digital command, the flipside of these visions involved dystopian fears about buttons and button pushers as immoral or unethical. As button pushers experimented with the extent of their fingers' reach—and what kinds of effects they could generate—they had also begun to think differently about how to command machines to do their bidding. Therefore, chapter 5, "We Do the Rest," studies how the act of pushing a button served as a bridge between human and machine labor, as button pushers could "call" electricity into action through consumer technologies meant for pleasure and instant gratification at one's fingertips. Yet concerns about "automaticity" and the consequences of doing away with human help also pervaded discussions about push-button interactions. Continuing to think about how the electrical industry positioned button pushing as a nonthreatening, magical solution to increase consumers' adoption of electricity, chapter 6, "Let There Be Light," takes a close look at push-button lighting. The chapter considers various concerns related to electrification that included concealment of electrical mechanisms, groping in darkness, and the efforts of electricians and homeowners to manage who could access light.

Part III, "Imagining Digital Command," charts negotiations over what buttons could and should do for their users as the concept of pushing a button became increasingly prevalent. Investigating actual uses and futuristic prophecies about life in a push-button age, this section unpacks enduring fantasies and fears associated with button pushers and button pushing related to work and pleasure. Chapter 7, "What's a Button Good For?"

returns to the subject of push-button communication discussed in Part I and tracks how thinking evolved about one-way summonses. In particular, it notes that although many critiqued ringing as a practice for getting servants' and workers' attention and began to prefer telephones and intercom systems instead, buttons were perceived as unparalleled assets for their reachability and effortlessness in situations related to panic and emergency (fire, burglary, automobility). Building on this investigation of how various user groups thought about the utility and purpose of push buttons, chapter 8, "Anyone Can Push a Button," follows continued deliberations to "groom" button pushers, particularly in the workplace, to make their bodily movements streamlined and efficient. However, these endeavors ran into strong opposition from workers about the tenets of digital command as "nonwork," with fears that previously skilled (manual) workers would find themselves replaced by digital button pushers. Chapter 9, "Push for Your Pleasure," concludes the section by examining advertisers' and electricians' persistent focus on electricity as a willing "genie" or "servant" made available at a push. The chapter identifies problems that began to occur when consumers took buttons (and therefore electricity) for granted without understanding or worrying about what made their pushes possible.

Last, the conclusion tracks a brief history after electrification became more commonplace by the 1920s and 1930s. It gives special attention to the push-button "craze" of the 1950s and 1960s Cold War era, demonstrating how societal definitions of pleasure and panic grew increasingly intertwined. Then, turning to the complicated present moment, it examines expanded definitions of what a "digital" button can do, how a hand works with its digits in digital environments, and the ways in which buttons have been constructed to hide the messiness of labor and technical infrastructures. Most important, the conclusion emphasizes the remarkable staying power of push buttons for more than 100 years in spite of persistent efforts to deride button pushers as lazy, unskilled, privileged, and even dangerous. In this friction between buttons as deliverers of pleasure and as harbingers of doom, it becomes evident that buttons continue to crystallize enduring societal hopes and fears about "easy" technological solutions to all manner of problems.

I   You Rang?

1 Setting the Stage

In as far as what really matters in life, pain, pleasure, danger, death, it is generally constituted by or caused by events occurring in the medium of touch. ... It is in the tactile world that we "live" in as far as it is mainly events located in it that can please, hurt or kill us.[1]

The act of pushing a button provided a solution to a number of thorny problems. For one, a host of individuals and groups from inventors and electricians to factory owners and domestic engineering specialists were focused on the problem of reducing manual effort. The minor act of pushing buttons to actuate machines could, theoretically, make muscular work unnecessary. They sought to employ technologies that could reduce the intensity of hands' labor, thereby redistributing action from the whole hand or hands to but a single finger or even fingertip and from vigorous exertion to a "mere touch." Indeed, the phrase "by the mere touch of a finger on a button" circulated across nearly every industry, becoming cliché in its regularity.[2] Sweeping changes in ideas about using one's digits emphasized that "no mechanical force is therefore necessary to be exerted," and instead "the mere touch of a key, register, pedal, or finger-button" could provide easy manipulation of almost any device.[3] "Mere touch" seemed to work as if by magic, for "by the mere touch of the finger on a button whole cities are aroused from their slumbers into a blaze of light and life before us like fairylands," one electricity enthusiast wrote.[4] Whether in factories, offices, or consumer spaces, designs for control with the smallest touch of a button promised that a newly defined hand could effortlessly command the object of its desires. To fulfill this promise required constructing a category of handwork that involved operation without effort.

Although push buttons attracted attention as an ideal mechanism for reducing hand labor, efforts to convert many hand movements to a single one extended beyond buttons specifically. Indeed, in the pursuit of nineteenth-century comfort through mechanization, many innovations involved "a single abrupt movement of the hand [that] triggers a process of many steps" in which "a touch of the finger sufficed to fix an event for an unlimited period of time."[5] Technologies like matches to start a fire or the switches used on a telephone switchboard also demonstrated a societal interest in rethinking how hands could carry out everyday tasks swiftly and efficiently, and engineers boasted about replacing difficult, hand-operated machinery with easy-to-use electrical machinery—cranes and hoists filled in for conveyors and chutes, while push buttons took the place of "laborious hand-actuated levers" and various kinds of "pulls."[6]

Yet this rhetoric of effortless machine interactions had long created uneasiness about what humans—and their hands—were supposed to do in a machine age. In treatises on laboring with machines, a perception of hand practices as atrophied had often figured importantly as evidence of human beings' alienation from production. Well before push buttons achieved mainstream usage, Marx (1848) worried in *The Communist Manifesto* that, "Owing to the extensive use of machinery and to division of labour, the work of the proletarians has lost all individual character, and consequently all charm for the workman. He becomes an appendage of the machine, and it is only the most simple, most monotonous, and most easily acquired knack, that is required of him."[7] Marx's reference to "appendage" demonstrated a prominent fear that machines had made human beings only useless extremities, ready to provide unskilled input without authentic engagement in the production process. This opinion, routinely echoed at the turn of the twentieth century and beyond, imagined that users of automatic machines would be reduced to "spectators."[8] Concerns that machines would minimize hand gestures to rote, meaningless activities—or replace them altogether—characterized numerous discussions about mechanized labor. However, in actuality, the finger that pushed the button might be characterized by power, influence, and engagement, or by impotence, inactivity, and ridicule. These determinations only grew out of associations and links between people and machines and never independently from them.[9]

Despite the fact that hands pushed buttons, just as they pulled ropes or gripped hammers, the act of button pushing rhetorically and physically disassociated from manual labor as a less demanding—and thus somehow inauthentic—hand gesture to those accustomed to working with their hands. Workers had to negotiate what push buttons meant for their industries and for the kinds of work performed in spaces ranging from offices to factories. These worries over manual labor fit within changes occurring to the American labor force, which included a shift from agriculture to industry, greater separation between producers and means of production, and declining control over the way work was conducted.[10] Using technological mechanisms to amplify human bodies for work particularly accelerated at this time period as human beings increasingly gave tasks over to and worked alongside machines.[11]

In line with these negotiations prompted by electrification and industrialization, proponents of electrical machines strove to redefine "work" both in and out of traditional workplaces. In the move toward streamlining workplace activities through minimal finger touches, disparities in terms of whose hands "worked" and whose hands "directed" stoked tensions among employees of various ranks. Laborers who spent their days doing physical activity expressed disdain for button pushers because they defined "work" as physical exertion, where those who sat idly by at their desks dispensing orders—exercising nothing but their fingers—did not qualify as legitimate workers in the eyes of others.[12] Rhetoric of this kind reflected the agenda of the scientific management movement, in which workers were taught to calibrate their hands with buttons that would enable them to perform "efficiently" with machines. To this end, scientists and efficiency experts rigorously documented the movements of bodies in factories and cataloged workers' efforts in experimental laboratories to achieve optimal button pushing. Managers, too, enthusiastically invested in scientific management principles and applied electrical solutions to minimize handwork and effort, pushing buttons to enforce discipline, control, and rationalized movement of bodies. Yet beyond these idealized views of pushing a button as an antidote to laboring hands, conflicts over how to define hands and what they could (or should) do reflected growing anxiety about mechanization. As workers pondered the definition of "work" in the industrial era, pushbutton control threatened previously stable understandings of labor.

As previously discussed, the act of pushing a button precipitated tensions between manual laborers and a growing class of "digital commanders," who managed others with their fingers rather than getting their hands dirty. Although we might commonly think of hands as tools of physical labor—the "hired hand," the "factory hand," the "field hand," and so on—a more nuanced exploration of hands at this time period reveals complexity and uncertainty about hands' roles.[13] Pressures to become "digital," to use one's hands to manipulate and control machines with an effortless finger touch, destabilized long-held beliefs about the hand as a modality for work, play, and everything in between. Buttons functioned as powerful scapegoats and symbols that emblematized a particular historical moment in which easy technological experiences were especially valorized and feared. To this end, push-button practices garnered so much attention because of broader shifts in meaning around "work," "manual labor," and "human touch" occurring across industries.

Although the notion of reducing bodily effort was part and parcel of the concept of pushing a button, other factors made this hand practice desirable, too. In a period of consternation about electrification, societal concerns turned toward the problem that electricity, an invisible force, evaded the senses. Writings on electricity often referred to it as "truly invisible" and "insensible in every shape or form."[14] Although close at hand, one could only verify its existence through effects. However, those effects often came at a price because touching an electrified surface could cause physical harm. So noted James W. Steele (1892) in his treatise on electricity: "Docile as [electricity] may seem ... it remains shadowy, mysterious, impalpable, intangible, dangerous. It is its own avenger of the daring ingenuity that has controlled it. Touch it, and you die."[15] A society grappling with electrification widely debated electricity's "shocking" nature, both fearing and admiring the power of electric shocks. Medical applications of electricity promised to ameliorate all kinds of ills, while reports of electrical accidents filled up the pages of newspapers and magazines as evidence of the energy's unsafe nature.[16]

Concerns over touch as a destructive force certainly predated the advent of push buttons; in fact, touching throughout history has often connoted potency and danger, sometimes referring to an insipid form of communication or harmful transfer in the case of bodily transmission of illness. To be "touched" by illness or to communicate by "contact" implied that touch could constitute a dangerous act; the word "contagion" came from the Latin root for "touching."[17] Discussions of infection and hygiene practices focused on "direct personal contact."[18] Feeling the breath of an infected person, too, constituted a form of "communication."[19] The closeness of bodies—and their intermingling—represented a pernicious touch whereby disease could move from one person to another. According to a German physician on the subject of antiseptics in surgery (1881), "One touch of a wound with a finger which is not surgically pure may lead to a fatal result."[20] As discussions about germs reached new heights in the twentieth century, scientists and physicians began emphasizing that even bodies touching surfaces—and not each other directly—could communicate germs. Indeed, as a guide for embalmers and sanitarians warned (1913), "One single touch of the finger, moistened with saliva to aid in turning the pages of a book, might contain over 5,000 germs."[21] Touching not just anyone—but also anything—could bring about catastrophe, and this potency generated fear about fingertips' potentialities.

Much as with germs, the invisible and intangible quality of electricity created friction with a perception that, as a sense, touch was viewed as a primary way of experiencing and confirming one's reality. Many believed that "through touch we largely acquire our ultimate notions of the externality, extension, solidarity, and permanence of objects, which are so much more 'tangible' than the reports given by the other senses."[22] Indeed, psychologist George Wallace Neet (1906) wrote, "whatever seems real to the touch has met the supreme test of reality. 'Let me take hold of it,' is our demand when we distrust our other senses."[23] How, then, to manage the intangibility of electricity with a desire to touch? To "take hold" of electricity constituted a kind of faith, but it also meant interacting with it free from fear of harm, to wrangle it so it might perform according to the rhythms and regularities of everyday life.

This dichotomy between tangibility (the desire to touch so as to make real) and intangibility (invisible and dangerous to the touch) is notable: humans could not palpate electricity; it existed in the shadows. As Steele surmised, one could die by virtue of a touch. Yet at the same time, a strong desire existed to harness and come into contact with electricity. To deal with this duality, electricians and early users relied on and experimented with ways to manage electricity by strategically concealing it. By "covering up" electricity—burying wires inside walls and hiding it behind push buttons—it could thrill, surprise, and delight without threatening. The act of button pushing emerged as a technique of protection, where one could conjure anyone or anything she desired at a safe remove from the imagined dangers of the Industrial Era. Buttons made electricity simultaneously real and yet magically and safely concealed. As a 2014 study on elevator push buttons has suggested, this concealment made buttons potent because they "sever[ed] the visible connection between cause and effect" by virtue of the fact that "the entire mechanism—electrical connections, control apparatus, motor—vanished behind the scenes. Only the push button remained visible on the surface like some last vestige and seemed to be responsible for the whole spectacle of motion all by itself."[24] Efforts to make buttons visible and tangible while tucking everything else away thus could act as a coping mechanism to make the untouchable touchable. In fact, by the 1900s, buttons' role in acting as the harmonious "face" of electricity received specific mention: "The button idea is a beneficial one, it not only embodies the principle of efficiency, celerity, snap and perfection, but it acquaints the world with the thought, that the old system, with everything exposed in its raw and operative state, is unnecessary. Food is not served in kitchens. Show rooms are separated from factories. The office is distinct from the place of the producers."[25] In this regard, a preference for buttons stemmed in part from their ability to hide machines' "raw" and messy parts.

Just as buttons could cover up electricity so it would turn on and off at will, hidden away until needed, so too did push buttons serve the function of managing the presence and absence and comings and goings of people. As noted, architecture—informed by new social practices—began to change: homes featured separate servant quarters; managers worked in offices away from their employees; apartments grew larger and taller, farther from the happenings of the street; automobiles put drivers at a remove from other drivers and from pedestrians. These shifts meant that people, no longer within earshot of one another, needed a strategy for garnering attention and making people strategically present or out of sight. Designers of homes, transportation, amusements, consumer products, and so on viewed push buttons as signaling tools that could facilitate these spatial rearrangements.

The more that people required greater isolation from each other, and from the machines they used, the more they required systems to overcome, manage, or enforce this distance. Efforts to extend the human hand's reach—to get "in touch"—fit into broader changes in transportation, communication, and control, which involved "the reordering of distance, the overcoming of spatial boundaries, the shortening of time-horizons, and the ability to link distant populations in a more immediate and intense matter."[26] This tension between proximity and distance raised important questions about the ways that hands should navigate newly electrified environments, and it also muddled long-held beliefs about touch as a proximate sense. Other senses seemed to travel across distance by virtue of electrical technologies. As Edward Bellamy (1897) suggested of this era, "You stay at home and send your eyes and ears abroad to see and hear for you."[27] The sense of touch, however, posed a greater challenge in terms of its portability. Physicians and scientists commonly confirmed that touch did not "extend beyond the reach of the arm" and relied on direct contact between two bodies.[28] Notably, it has been argued, "of all the senses, touch is the most resistant to being made into a medium of recording or transmission. It remains stubbornly wed to the proximate; indeed, with taste, it is the only sense that has no remote capacity."[29] Yet technologies like push buttons that could spur action over distance with a finger seemed to defy this logic. Indeed, author George T. Lemmon (1899) poetically wrote, "Our fingers have grown immensely longer. We touch each other from vast distances these days."[30] Although Lemmon might have referred to a kind of metaphorical "touch" based more on communication than physiology, his words spoke importantly to a burgeoning relationship among hands, touch, reach, and distant effects that once seemed impossible.

Within these contexts, "pushes" (another name for buttons) came to serve as the most familiar kind of control mechanism for domestic and commercial purposes in the United States.[31] The type of switch employed mattered on a number of levels, from its technical capabilities and cost to its aesthetics and correspondence to the surrounding environment. For example, Frank Eugene Kidder recommended to architects and builders that home designers should choose the button option when "a neat appearance is desirable."[32] He noted that electricians should implement snap switches (which looked similar to today's oven dial and were operated by the turn of a wrist) when looks did not matter because they were inexpensive but less pleasing to the eye. Thus, snap switches constituted an appropriate choice in vestibules or hallways, but highly trafficked areas such as sitting rooms, dining rooms, and parlors should feature push buttons.[33] Meanwhile, knife switches acted primarily as circuit breakers, meant more for industrial use and higher voltage.[34] As Kidder noted, buttons became popular for everyday use not because they worked better than other switches but because they could be disguised as not-switches; aesthetically, they blended best with their surroundings so as to appear unobtrusive. Buttons came in all shapes, sizes, and degrees of expense; the "common variety" wood ones used walnut, rosewood, oak, maple, or mahogany. Compared with those more ornate buttons made from metals like brass, wood buttons earned a reputation as "cheap and ugly."[35] Due to buttons' many variations, experts and homeowners chose materials for these buttons depending on their purpose and degree of visibility in the home.

In the early 1880s, few electric buttons existed because few electric devices were available to the general population. An 1882 catalog, for example, offered consumers three push button options: a pear-shaped push button ("To be attached to Electric Bell"), a compound push button (a panel with three buttons designed for office use so that managers could buzz a cashier or assistant), and a circular push button (in bronze, nickel, or wood) for "insert[ing] in desks or other furniture."[36] These buttons ranged from 75 cents to $2.50 a piece in cost and occupied but half a page in a catalog of more than 100 pages. Two years later, the same catalog had expanded its offerings to one full page of buttons for purchase, most with the same technical features, but providing larger, more detailed illustrations.[37] By the early twentieth century, more than 50 different designs of push buttons existed at a fraction of their previous cost.[38] Consumers could purchase buttons that clamped to dining room tables and embedded in floorboards for easy pressing by hand or foot; they could obtain buttons with lettering, numbering, and intricate decorations; and they could choose buttons that hung from cords, illuminated, and that made a variety of sounds. Ranging from plain to incredibly ornate, push buttons in the early twentieth century had evolved into inexpensive, desirable, and multifaceted electrical accessories (see figure 1.1).[39]

As more consumers began to consider electrification, a host of electrical supply companies started producing push buttons. Some of the most prominent manufacturers included Cutler-Hammer Manufacturing Company of Milwaukee, Hart-Hegeman Manufacturing Company of Connecticut, the Perkins Electric Switch Manufacturing Company of Connecticut, and General

CATALOGUE AND PRICE LIST. 15.

PEAR-SHAPED PUSH-BUTTON.
This Push-Button requires but a very slight pressure of the finger to ring the bell. They are invaluable for the sick-room; can be laid directly on the bed, or hung up when not in use.

CIRCULAR PUSH-BUTTON.

NICKEL PUSH.

BRONZE PUSH No. 1

BRONZE PUSH No. 2,

FOOT-PUSH FOR FLOOR.
Pin is loose and can be removed.

DESK PUSH

NEW STYLE SIGNAL OR TAP KEY.
[Closed or open circuit.]
A very neat and substantial key for Telephone Calls, Electric Bells, etc.

Electric of New York. Alongside these notable producers, dozens
of others also entered the market with an eye toward capitalizing
on the growing demand for electrical control mechanisms.[40]

Buttons worked in a variety of ways, and how fingers pushed
them often aligned with how people perceived their purpose and
effect. Most buttons functioned via "momentary" action. This
momentary push came out of key culture; as with the telegraph
key, the piano key, and the typewriter key, one could only get
sound, signal, or effect when one's finger touched the button.
The camera button or electric bell button worked in the same
way: a snap of the shutter or ring of the bell tied to the action of
the finger. Stringing together a set of sounds, patterns, lights, and
so on became important because meaning was derived from how
many pushes a finger made and how the pusher arranged her
pushes according to certain patterns and rhythms. In this case,
buttons defaulted to a state of "off" would spring to "on" upon
pushing and then return to "off" again. Alternatively, "continuous" buttons could retain their state after an initial finger pressing. This functionality meant that buttons "automatically" took
over the work of state maintenance—holding a button in an on
position for an extended period of time—and the only time a
pusher interacted with a button again was to turn it off; therefore, the finger only intervened to begin and end something.
Yet a third form of button pushing utilized buttons in a discrete

Figure 1.1
Catalog advertising a variety of push buttons available for purchase.
Source: Patrick, Carter & Wilkins, *Catalogue of Annunciators, Alarms
and Electrical House Goods*, 1909. Image courtesy of the Warshaw Col-
lection of Business Americana—Electricity, Archives Center, National
Museum of American History, Smithsonian Institution.

manner, much like a trigger.[41] To use a button as a kind of trigger meant that the finger could start a process or action; once the process began, it would no longer remain under the control of the operator's hand. Trigger buttons existed more in the cultural imagination than in everyday life, as in fears about push-button warfare, and they presented a weighty proposition: as the process of state maintenance became increasingly autonomous from the finger, one could not undo what was done. There was no "off" switch and only one push: an irrevocable one.

In the spaces where digital hand practices were implemented, workers coped with increasing physical distance, new production processes, and bureaucratic measures that destabilized traditional working environments. Efficiency experts touted the benefits of improved reaction time and enthused about hands that could do more by virtue of their machine counterparts. Similarly, advertisers boasted to homeowners about the merits of button pushing that reduced the need for hand strength or special skills. A glorified vision of a single push meant to imagine a world in which "huge machinery is started and stopped through it by one-finger power."[42] Yet the mythical button that eased the hand's burden, required only a finger, and functioned without problems did not really exist. Just as no universal hand pushed buttons, buttons did not operate seamlessly and neutrally.

2 Ringing for Service

Some persons object to pushes and pressels; they like to have something to pull as in the ordinary bell-pull of the old system. There are others who have become habituated to the bell-pull, and cannot take comfortably to the order of new things.[1]

The practice of ringing bells served an important function in the late nineteenth and early twentieth centuries, and in fact the bell acted as "a proxy for all sorts of relations" between people.[2] To understand how buttons began to achieve prominence as communication and control mechanisms requires understanding bells and bell ringing as a popular method of signaling, warning, garnering attention, and making demands. One might find push-button bells in a variety of places, but they first achieved prominence as fire-alarm mechanisms. In 1870, inventor Edwin Rogers designed a fire-alarm repeater and included a push button in his designs. Deemed the "inventor of the electric push button" years later in his obituary (the only fleeting mention of buttons having any one "inventor" at all), Rogers identified a need for instant action and reaction—and signals that could replicate over distance—to engage the appropriate parties in fire response.[3] Rogers' repeater made it possible for an alarm triggered at the location of one box on the street to strike multiple bells and gongs throughout the system, sending an alert without the aid of a central operator.[4] A finger on a button worked to make these signals rapid and communicable with little human intervention, and this repeater technology offered particular value to small towns that could not afford a central office system.[5] Observers took note of the "electric communication between all quarters of towns, and between many houses, and numerous fire-stations ... everywhere" made possible by a finger's pressure upon a button.[6] Large buildings frequently employed push-button signals given the potential for fire or other kinds of emergencies in those spaces. Public theaters, for example, featured push buttons in glass cases to trigger alarms in case of fire, a kind of early panic button that equated the technology with warning and disaster.[7] Theater fires were all too common: by 1878, fires had destroyed 516 facilities, prompting routine installation of fire-alarm boxes in most of these spaces.[8]

This case of pushing buttons in emergency situations demonstrated the potential potency of exerting minimal effort to signal across distance, to command responders to action. Such a concept took root from the factory floor to the bank, where call bells facilitated information transfer among employers, employees, customers, and others in ways that did not (theoretically) disrupt the flow of work.[9] Of all these locales, hotels were the primary early adopters who integrated bell outfits into every part of their operations, and these establishments pioneered many of the push-button advancements that ultimately trickled down into other spheres.[10] Renowned hotels, such as the Palace Hotel in San Francisco and other upscale American hotels, began to provide the push-button feature to patrons in the mid-1800s.[11] Large hotels required elaborate electrical systems to make push-button connections among their various parts. A hotel with 500 rooms across six to ten stories might require 1,200 electrical appliances, including bells, buttons, and speaking tubes.[12] Within these systems, many hotels employed an "annunciator," a kind of pre-intercom push-button system that linked up rooms, the central office, halls for chambermaids, engine rooms, the laundry, and any other places where proprietors might want communication.[13]

In its simplest form, the annunciator functioned by enabling an individual to press a push button that would close the electrical circuit; this push triggered a "drop" in the hotel office (the drop was usually a card, needle with a pointer, or flag moved by an electromagnet), that would indicate which room number made the call while also activating a bell to ring (see figure 2.1).[14] The attendant could then visit the guest's room in question to receive orders and carry out a task as needed. Inventors expanded on this design so that guests could make specific requests rather than waiting for a bellboy to visit the room (see figure 2.2). The engineer F. Benedict Herzog, for example, created the popular "Herzog Telesme," an annunciator that enabled the guest to ask for anything from a bottle of champagne, to a bottle of ink, to one's luggage. The guest would move a pointer on a dial to the object she required in her room and press a button, transmitting a "signal that indicates on the board at the office just what is called for."[15] In addition to its practical function, which aimed to make the button push more intelligible, the Telesme played on well-known psychology that people liked to press buttons for instant gratification to summon one's desires with the touch of a finger. Other systems took a telegraphic approach, which, like the bell-pull system that preceded it, required the pusher to indicate her desire with an assigned number of pushes.

Figure 2.1
Annunciator technologies used push buttons to enable signaling in large buildings, such as apartments or hotels, when users needed a communication mechanism across distance.

Source: Novelty Electric Company, *Illustrated Catalogue and Price List* (1899): 16. Image courtesy of the Warshaw Collection of Business Americana—Electricity, Archives Center, National Museum of American History, Smithsonian Institution.

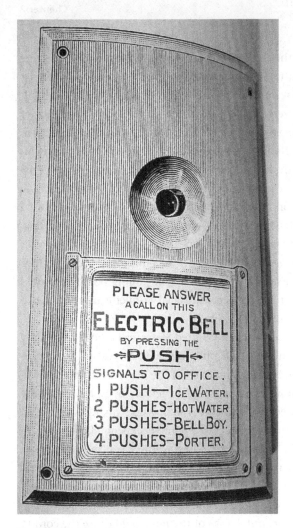

Figure 2.2
Hotels employed annunciator and call bell technologies to make guests'
requests known to staff with the press of a button.
Source: Frank H. Stewart Electric Co., Catalog, n.d. Image courtesy of
the Warshaw Collection of Business Americana—Electricity, Archives
Center, National Museum of American History, Smithsonian Institution.

Although the notion of pushing a button took hold in large buildings in the mid-nineteenth century, in domestic contexts people did not push to ring bells; rather, they pulled. Mechanical bell pulls were most common, which required a strong tug of a cord often made of silk. These pulls reflected Victorian fashion, and women often hand-stitched them. They reflected "individuality ... in both design and execution" and were an "indispensable item in the catalogue of feminine accomplishments."[16] The communication of messages via bell pulls sometimes required only one pull, whereas other systems functioned more like telegraphing a message, in that one pull signaled for a servant's attendance, two requested fuel, three pulls asked for light, four for water, and so on.[17] The practice of pulling bells substituted for in-person requests because social etiquette deemed it bad form to "call out" to servants.[18]

Although popular in the United States and abroad, especially in well-to-do homes, bell pulls garnered a great deal of negative attention. Some complained of the physical effort and awkward hand movements required to actuate bells, as in the case of one user, George Edwinson (1883), who remarked, "What an immense amount of muscular energy is expended every day in pulling bells!"[19] He went on to ask, "Can't you men of science devise some better method of summoning our servants and making them acquainted with the fact that some person desires their presence?"[20] Users indicted the difficulty of pulling bells properly because horizontal jerks of the pull could cause the cord to break or the decorative tassel to come off in one's hand.[21] Stories about using bell pulls went as far as to suggest that only an "extremely athletic man," using two hands and firmly planted feet, could cause a bell pull to move.[22] In other instances, bell pullers lamented that bell ropes routinely went out of order so that they did nothing at all, either broken or hanging limp and loose.[23]

Early doorbells, which also functioned by a pull, usually of a knob, garnered similarly unsavory reports. According to one experienced user, "You pull its clammy knob out of its rusty socket, let it go, and listen. No sound. You pull it a little further out, and listen again. Less sound. You lug it forcibly out of its socket to its fullest length, and let it fly like a catapult, and there is a slight tinkle-tinkle in the distant down below."[24] Much to the author's chagrin, only after one's "fourth and final paroxysm" at the pull would the caller get a response.[25] Known for the forceful hand movements required to activate them, bell pulls married a forceful tug with a forceful demand for a servant's or attendant's presence.

Those homes without knobs for doorbells featured doorknockers made of heavy metals like brass, and they often displayed ornate carvings or embellishments. Like bell pulls, they usually reflected the character of the house's architectural style. For some, this unique construction recommended pulls for the way they interacted with individual hands. In fact, a fiction story told from the perspective of a doorbell illustrated this viewpoint. According to the doorbell, "I assure you we can distinguish between rough and kindly touches, and aggressive pulls and circumspect pulls, and insolence and courtesy," suggesting that, although pulls might generate frustration, they also related to a wide variety of distinct human touches.[26] As with the unique and handcrafted nature of bell-pull material, how people used their hands to knock on doors related to hand strength as well as personality.

Beginning in the 1870s, push buttons to actuate bells rose to prominence with the advent of electricity, which necessitated new ways of arranging these signaling devices as well as new methods for triggering them to ring. Electric bell outfits were operated by a battery; rather than a pull, the ringer would press a wired push button that would close an electrical circuit.[27] In contrast to the bell pull, the electric button limited what the user could do with her hands based on the palette of sounds available. Buttons did not ring harder or louder if pressed with more force or intensity; therefore, their use created the conditions for a new sensory, kinesthetic, and aesthetic experience. Indeed, descriptions of push-button bells emphasized that, "A gentle pressure upon a small button effects all that is required. The electric force that rings the bell steals noiselessly along the wire, there is no sound, there is no strain; but the bell gives forth its warning sound as though it were rung by the stout arm of some invisible sprite."[28] This "gentle pressure" of the finger and electric force noiselessly carrying along wires differed quite significantly from prior descriptions of actuating bells or knockers, which required rigorous and athletic movements and generated a host of sounds that corresponded with the hand's individual knock or pull. In this regard, pushes presented a new model for calling someone to action. Noted one user of another popular style bell, "Bells you bang on tables hurt your fingers, and invoke bad language—but not the servant. I never knew a good-tempered household that used bang-bells at the table. They are a standing invitation to violence, whereas the press-bell pleads for gentleness and restraint."[29] Some equated the minimal or "gentle" hand intervention required for pushing with a more polite, discreet, and therefore gentle form of managing household staff.

Reports varied widely on whether pushing a button to operate an electric bell constituted an improvement or a hindrance compared with the bell pull. Some saw the act of pushing as representative of a new modern era, as in the words of one writer: "The knocker is out of date. A pull is passé. It now takes an electrified punch."[30] To these enthusiastic adopters, bell pulls were viewed as "elderly," and "grandma and grandpa can't quite reconcile themselves to this modern method of pressing a button, but would feel happier if they could be permitted to pull something."[31] To this end, a number of observers remarked that push-button bells in public places often accompanied signs marked "Bell" and "Push" to move along "the slowness of people to acquire new habits."[32] Those who lamented the arrival of push buttons criticized the "press-the-button-fiend" and lamented, "With the brusqueness characteristic of the times we are instructed to PUSH this modern startler."[33] Comments like this one reflected a perspective that pushing constituted more than just a change in mechanism; it also symbolized a shift in philosophy regarding the forcefulness, speed, and "pushiness" of the electrification and industrialization era.

As a result, remaking the act of pushing as a nonforceful one—converting pushiness into "mere touch"—constituted one of the primary projects of this time period. To do so, in part, involved redefining pulling and pushing metaphorically and philosophically beyond the hand movement. The "pushing" person typically connoted someone self-made and industrious. Indeed, "The 'push' individual is like the bird which has graduated from the nest and is able to forage on its own account; the 'pull' person is like the weakling in the nest which requires constant feeding in order to prevent it from starving."[34] Similarly, an editorial on the social climber who could advance her position from one social status to the next (1880) noted that, "Pushing is carried on with very unequal degrees of skill. There is a clumsy and an adroit kind. Just as a man may physically push his way through a crowd in a rough fashion, calling everybody's attention to his exertions, while another will get through quite as effectively without any disturbance or appearance of effort."[35] In an interesting turn, the "pusher"—whether pushing to achieve a better social position or actuate a push button—achieved success when exerting as little effort as possible, gently charming her associates.

Although electricity served as a major topic of curiosity and consternation in the late nineteenth century, electric bells received little attention as novelties due to their simplicity; they were one of the most "familiar examples" of electricity and became popular before widespread public use of the telegraph and even helped to make that device possible.[36] According to an encyclopedia entry on the topic (1880), "The ringing of bells is not a recent application of electricity, but it is only a few years since electric bells have been placed in many public and private buildings instead of the well-known bell hanging arrangement with wires and cranks. ... In every room which communicates with a bell there is a 'press-button' or little spring by which the current of electricity is put off or on as we may wish."[37] Eventually, the push button was used for ringing a bell both indoors and outside and was perceived as "too common to require description," according to John Henry Pepper (1881), and "so well known that we need not describe it," in the words of E. Hospitalier and C. J. Wharton (1889) in their book *Domestic Electricity for Amateurs*.[38] However, despite a perception of push buttons' ordinariness—especially for those in the electrical industry— early push-button communication, especially in domestic contexts, occurred primarily between those people and in those spaces already enjoying the privileges of wealth before in-home telephones became common.[39] Early users were also commonly urban users, and although these "city dwellers ha[d] come to believe in the prevalence of the electric button," they constituted but a mere fraction of the larger population.[40] This observation is unsurprising given that most homes were not wired for electricity until well into the twentieth century.[41] Indeed, many homes at the turn of the twentieth century did not even have doorbells, such as tenements in New York, where by 1890 the doorbell was "practically an unknown institution" (see figure 2.3).[42] According to educator Jennie Darlington (1889), "I have known persons who were afraid to have an electric bell in the house, and have been told that such cases are common. These persons are even afraid to ring an electric door-bell."[43]

Middle- to upper-class homes and large city apartment buildings were commonly outfitted with push-button systems that could make interactions less laborious. As in hotels, annunciators were often installed in apartment houses to help individuals carry out tasks from many stories above, such as summoning a messenger boy, calling a carriage, or requesting an object from the cellar. Using an electric button on the wall in each room, someone could push a button and turn a marble handle to indicate the service she desired. Noted an article on the subject, "In a moment or two, a servant knocks at the door with the thing you called for. ... It is an odd way to live, far up in the air above the house-tops."[44] As early as 1892, thousands of people in New York "wonder[ed] how they ever got along without their annunciators, telephones, bell-calls, door-openers and elevators."[45] Servants in these buildings used such systems to carry out the business of the day. Upon receiving a call from a peddler, for example, a maid could choose to respond at a distance by using a speaking tube to get more information, trigger a door to unlock

ORNAMENTAL

BRONZE FRONT DOOR PUSH.

NEW DESIGN.

Having cap made to unscrew in same way as Wooden Push Button, thus giving access to the springs without removing the plate.

PRICE.

Oblong $2.00
Round 1.00

OBLONG.

ROUND.

Figure 2.3
Buttons with ornate designs and patterns, especially used in doorbells, provided an aesthetically appealing appearance to wealthy homeowners and their guests.
Source: J.H. Bunnell & Co., *Illustrated Catalogue and Price List of Telegraphic, Electrical & Telephone Supplies, No. 9* (January 1888): 130. Image courtesy of the Warshaw Collection of Business Americana—Electricity, Archives Center, National Museum of American History, Smithsonian Institution.

for his entry, or wait quietly for the peddler to go away.[46] City children, too, would often have exposure to "a few simple phenomena; a push of a button—a bell is rung; another push—a door is unlocked; another push—a light appears."[47] They would come to have familiarity with the fact that "the modern apartment is a complicated structure operated by buttons."[48] A rapidly changing cityscape necessitated new devices like the push button to manage information and communication.

In electrified homes, large spaces demanded a means for managing staff, services, and the comings and goings of dwellers and visitors. Electrical catalogs, product manuals, housekeepers' guides, electricians' plans, and numerous other instructional documents weighed in on various methods for constructing and using buttons to facilitate communication in domestic spaces.[49] Electrical materials—bells, push buttons, batteries, and wire—had also become affordable to the point that one could realistically achieve home installations and fixes.[50] Homeowners expected servants to respond to bell calls as part of their daily responsibilities, which also included cleaning and bringing fresh water.[51] Some were even required to repair these bells and their associated buttons.[52]

As electric bells began to replace mechanical ones, electricians and homeowners advocated for doing away with unsightly wires that had previously accompanied bell pulls in favor of a harmonious push-button faceplate. Indeed, "At first it was usual to expose the wires to view along the walls and ceilings, even in the best houses," Edward and Francis Spon (1886) wrote in a manual to mechanics, "until the 'secret system' was introduced, which consists in carrying the wires and cranks and tubes and boxes concealed by the finishings of the walls."[53] Bells pulled by mechanical wires suffered from significant limitations in how bell hangers could install them in a house; wires ran in a straight line and often snapped, which required expensive repairs and pulling up of carpets and rugs to remedy the situation.[54] As a result, advocates for the "secret system" appealed not to home-owners' safety or security but rather to their aesthetic sensibili-ties. Bell hangers advocated for plans in which "not a single wire is seen in any room" to improve how a house looked.[55] This kind of pitch often appeared in advertisements for electrical products, as in the illustrated catalog of The Stout-Meadowcroft Company of New York (1885), where electricians assured that electrical apparatuses would cause "no inconvenience to the family or injury to the property as all wires are laid in concealed spaces by hands skilled in electric work."[56] In fact, the concealment strategy featured prominently in the electrical home, so that "the battery, which is enclosed in a neat box 9 x 15 inches, can be located in a closet or other convenient place, where a small space can be spared, as it emits no disagreeable or offensive odor, and it needs attention but once or two a year."[57]

Nevertheless, this goal presented a number of problems. One could encounter difficulties concealing wires by simply papering over them with wallpaper because this strategy created uneven lumps and attracted dust. By contrast, wires embedded at an ear-lier stage in plaster often caused inconvenience when it came to eventual repairs.[58] Patent descriptions often remarked at the difficulty of repairing embedded buttons and their wires, given that most designs privileged aesthetics over functionality and made their working parts hopelessly out of reach.[59] Beyond this problem of embedding, rats and mice were common enemies to electrical mechanisms and were viewed as but one of a num-ber of antagonists to properly working buttons, including short-circuited wires, slow make-and-break action, and the seeping in of foreign substances.[60] As a result, bells and lights notoriously malfunctioned, and when they didn't ring or stay on constantly, they often remained out of order for days when city residents didn't bother to have them fixed.[61] Even if someone wished to outsource push-button repair, she might find it difficult to contract the right help. Most bell hangers and electricians refused to waste their time with these minor electric mechanisms, preferring more elaborate wiring jobs.[62] Tension existed, then, between the promise of push-button magic—which relied on hiding the formerly "visible medium" of the wire and crank—and problems of accessibility and maintenance.[63]

Just as hiding wires away served an aesthetic function to make electricity more attractive in homes, so too did push-button designs offer a new opportunity to integrate control mechanisms into the surfaces of everyday life. Buttons could cover up a home's past technologies by erasing blemishes of the past. For instance, electricians recommended buttons in circumstances when homeowners chose to remove mechanical bells and replace them with electric ones. To get rid of the "ugly looking hole" left in the door, one technician recommended "naturally look[ing] to the electric push button as the most suitable thing to help us out."[64] Whereas bell pulls and their wires stuck out in plain view, electricians could insert push buttons inside walls, into floors, in desks, tables, and other surfaces so that they blended unobtrusively with their surroundings. By bringing these controls discreetly into living and working spaces, architects and electricians advocated for a new era in which "the controls of machines were grouped around the operator" so that the controller could access anything—or anyone—at an arm's length.[65] Calls for reachability often involved appeals to safety, convenience, or relaxation, and in the case of push buttons, they married the notion of a "single touch" and effortless engagement with the benefits of electrification. Electricians, architects, consumer product manufacturers, and others imagined users as ergonomically situated "armchair generals" (although neither the concepts of "ergonomics" or the "armchair general" existed yet in this specific language) who could comfortably control and communicate with the "mere touch" of a finger from anywhere. Such visions applied to specific users in positions of authority and to a certain class that could direct the movements and activities of others rather than undertake those movements themselves.

Along with using call bells and buttons to facilitate homeowner–servant relationships across distance, push-button bell systems also provided an opportunity for homeowners to protect their domiciles by acting as a form of enforcement, drawing on earlier uses of fire-alarm buttons. In popular magazines and newspapers, authors routinely described the alarm button as a comforting measure in one's home for those who feared intrusion. An article titled "Electricity in the Household" (1897) noted, "The class of persons who retain the traditional fear of the hidden burglar find great consolation in the secret push-button placed at the head of the bed and connected with an alarm at the nearest police station. ... The sensation of noiselessly touching the button and knowing that the more busily the gruesome visitor is engaged the more certain is his capture at the hands of the policemen who are hastening from the station, must be unique."[66] As with the discreet push affixed to the homeowner's table leg, the alarm button, secret and silent, performed a vital service for those who sought to enforce the boundaries of their homes by getting in touch.

Electrical supply companies sold burglar alarm apparatuses that offered "ABSOLUTE protection from Thieves" while also giving information about servants' comings and goings.[67] These outfits ranged in price from $2 to $25 depending on the materials in use.[68] The homeowner might even sleep with a push button under her pillow or hang a button from a cord above her bed to stun an intruder.[69] These various embedded surveillance tools created communication and feedback loops within domestic spaces, making them intelligible to their owners and disciplining the bodies that inhabited them.[70] Electricians often made detailed drawings and measurements of homes to provide complete alarm solutions that considered how these bodies would interact; they thought holistically about alarms as networks rather than independent mechanisms.[71] Early adopters of call bells and burglar alarms included noteworthy businessmen such as railroad magnates George Pullman and Perry H. Smith, who hired electricians to set up their Chicago homes.[72] The wealthier set had the means to install new electric alarm mechanisms, and they also possessed desirable goods that they perceived as needed protecting in the first place.

In public contexts, push buttons served as formidable defense mechanisms by facilitating quick action and reaction against burglars. Due to this fact, banks commonly installed push-button alarms, providing tellers and managers with protection from robbers who frequently assailed these businesses and even took the lives of employees.[73] Electrician and inventor Thomas Edison, an early and outspoken proponent of push buttons, encouraged bankers to hire electricians to install secret alarm buttons in all banks. Edison imagined that invisible buttons could make the difference between victimization and control. According to *The Electrical Journal* (1896), which interviewed the inventor, "Part of the floor about the entrance to the bank president's office could be arranged with metal plates so as to be charged with a high current of electricity when desired. The wires and other fixtures could be concealed beneath the carpet, or even in the flooring, without difficulty."[74] The article concluded, "The entire arrangement should be controlled by a button underneath the banker's desk. Even the button might be out of sight beneath a rug or carpet. This would be an immense advantage in an emergency."[75] The banker could alert police with a concealed button and a simple press. Far from fiction alone, Edison's vision in fact did come to fruition only a year later with an invention that allowed a banker under duress to push his foot on an emergency button that would send all money cabinets into lockdown and transmit a signal to call attention to the situation.[76] In this case, as with the homeowner demanding the presence of her servant, the button could facilitate a gentle and unassuming interaction that belied its forcefulness.

Tinkering communities often took interest in burglar alarms, which household inventors adapted simply by applying push-button bells to doors, windows, curtains, and other points of entry.[77] For the "householder with a scientific turn of mind," a temporary alarm could be constructed with wire, a battery cell, and a button at little cost.[78] In fact, the "average family junk box" contained the materials necessary to put an alarm in place.[79] Homemade projects were common as well because many alarms were not eligible to receive patents due to their ubiquity, as in the case for a product created by C. P. Jones of Colorado, who invented a safe connected to push buttons in the floor that would sound an alarm in a police station if activated.[80] Buttons' affordability, accessibility, and familiarity meant that many similar and iterative products circulated that were constructed by amateur tinkerers for individual use.

More broadly, authors of educational materials encouraged novices to take up push-button construction and repair for household projects and to think creatively about their uses.[81] These craft projects served to reinforce a "dominant ideology" because hobbyists usually created objects similar to those in the commercial world, and they "work[ed] in a socially prescribed way."[82] Late nineteenth- and early twentieth-century men and women often received instructions to rationalize their leisure time and were urged not to "waste" it by being productive outside of working hours; children were similarly taught at young ages not to succumb to laziness.[83] In this context, push-button tinkering fit within broader social norms that demonized "idleness" and valued work as a form of leisure.[84] Given push buttons' common association with opulence and laziness, users could reappropriate them by working with their hands rather than calling on someone else to do the work.

Numerous publications cropped up to help these tinkerers manage a growing set of electrical tools, but these were targeted primarily toward electrical experimenters and do-it-yourself types. Popular and academic texts often featured lessons on how buttons worked, describing electrical circuits and pushes for "on" and "off" in their first few pages. Titles such as *Domestic Electricity for Amateurs* (1889), *Everybody's Hand-book of Electricity* (1890), *Electricity in Daily Life* (1890), *Popular Electric Lighting* (1898), and many others provided detailed instructions and visual diagrams on buttons as electrical components.[85] Magazine articles, including "Electricity Applied to Household Affairs" (1893), "Electricity in the Household" (1897), and "Electricity as a Domestic" (1901), specifically addressed home repair for the mechanically inclined.[86] Although most of these texts identified men as their primary constituents, others appealed to women because push-button switches and bells often fell under women's jurisdiction as they appeared prominently in domestic spaces. Indeed, author Helena Higginbotham wrote to readers of *Good Housekeeping* (1905) that she had wired nine homes beside her own, and she articulated that any woman could easily gain this knowledge. Describing problems that someone might encounter when installing electric bells, Higginbotham gave special attention to push buttons and their potential misfires.[87] Rather than inaccessible, taken-for-granted mechanisms, buttons appeared in this text and others discussed earlier as simple technical objects that individuals of all ages should understand. These lessons viewed buttons as but one of many conduits that conveyed electricity and required education and exploration.

Many educators sought to inculcate the public so that they might adopt electricity broadly and without hesitation, given widespread reticence toward electrification.

Where it might prove difficult to persuade adults to use electricity, many educators believed that children, as first-generation button pushers, would not encounter buttons with the same distrust. Learning about a button's mechanisms could replace buttons' magical properties with a more practical rationality about electricity; to this end, push buttons acted as gateways to more broadly understanding electrical forces and indoctrinating the next generation of users.[88] This practical approach sought to demystify button mechanisms and make them mundane for children by linking them to lessons about how to "make" and "break" electrical circuits or connect an electric bell. For example, one educator concluded, "A little child, however, after observing a few simple experiments in electricity, probably thinks with perfect coolness when he touches an electric bell knob, 'Now I have closed the current.'" She noted, "We have not yet given the very smallest children the experiments in electricity, but mean to do so before long, and they are such as to remove the feeling of awe (as of unnatural agency) that many have for electricity."[89]

Such education often began in formal classroom settings for American youth, where educators taught students in elementary schools how to create electric bells, buzzers and buttons; schools considered construction of these household electric devices an important part of students' science curriculums. Even beyond the classroom, an incredible wealth of books and magazines targeted school-aged tinkerers, encouraging them to explore and understand their physical world, including the push buttons that animated their everyday environments. Buttons were relatively inexpensive and simple electrical mechanisms, meaning that novices could purchase or construct their own buttons.

Educators typically described boys as their target audience for lessons about electricity and magnetism because they believed that boys had a natural affinity for working with machines, and they assumed that boys gravitated toward and possessed a "remarkable ability for mechanical work" more so than they would take an interest in history or literature.[90] Views about gender roles led most professionals to agree that girls, too, could benefit from lessons on "domestic economy" and the simple devices such as a push button that one might operate in a home.[91] In educational journals, many teachers proposed a hands-on approach to electricity, where students could utilize the classroom as a laboratory for tinkering with objects they encountered outside of school: "We are not only acquainting our pupils with the great truths of science," wrote one educator, "but we are creating and fostering that most desirable and productive quality, the 'scientific habit of mind.'"[92] The notion of an active education, promoted by many such as John Dewey, envisioned the push button as a "situation" for children to encounter electricity and make a connection between a push of the finger and a ring of the bell.[93] Understanding how push buttons worked played an integral role, in educators' estimations, of thinking scientifically and rationally about everyday technologies.

Given the intrigue that surrounded push buttons and their ubiquity in so many contexts, teachers also viewed buttons and other electrical tools as a way to "hook" students to appreciate science more generally.[94] Paul H. Hanus, professor of education, asked his colleagues, "Why cannot we begin natural science with the study of the push button, the camera, the electric light or the lighting of a match? ... All studies should be taught with reference to their social significance."[95] Teacher Otis W. Caldwell similarly commented that concrete experiments—building and taking apart electric bells, examining telephones, telegraphs, and dynamos—enabled students to fully grasp their lessons.[96]

Beyond these lessons in classrooms about push buttons, a wide swath of print literature—popular newspapers, magazines, and books—also encouraged children to take an interest in constructing bells, buzzers, and buttons. Texts included *Questions and Answers about Electricity: A First Book for Students* (1892), *Real Things in Nature: A Reading Book of Science for American Boys and Girls* (1903), and *The Sciences: A Reading Book for Children* (1904), each of which included a section on buttons and their relationship to electricity.[97] One exemplar, *Things a Boy Should Know about Electricity* (1900) outlined various uses of push buttons for aspiring young male engineers, including affixing buttons to windows and doors for burglar alarms.[98] Similarly, an article in the *Atlanta Constitution* from the same year titled "A

Boy and a Bell" detailed how a boy should go about constructing his first bell for his mother. In a section called "The Push Button," the author noted that "the push button is so simple that the average boy can take two pieces of thin sheet brass, copper, or iron and make a temporary one in a few minutes."[99] Another instructional piece recommended that experimenters craft push buttons from inexpensive materials like aluminum and plaster of Paris.[100] Books in later years, such as *Harper's Electricity Book for Boys* (1907), told readers about push buttons' important role in making and breaking electrical connections, unpacking their logic while emphasizing how buttons related to the electrical circuit's overall functionality.[101] Although many texts targeted professional scientists and inventors, this primer and others focused on the ways that children could tangibly and meaningfully interact with electricity by beginning at the site of the push button.

On the surface, electricians and other inventors designed push buttons with accessibility in mind, surmising that users— whether young or old—should readily understand and possess the physical strength to operate push-button bells. Yet numerous anecdotes suggest that everyday experiences with buttons ran counter to these ideals. The tenet that anyone could push a button came with a host of baggage about what "anyone" would do with a push button and who exactly qualified as "anyone." Pushers routinely experienced confusion about how buttons worked and what effect they would produce, or they deliberately took advantage of buttons' technical simplicity for their own ends. Given a common perception that people should naturally know how to push buttons to achieve a desired result, to have an accident, mishap, or confusing experience with an electric button thus constituted an example of ignorance worth reporting (and perhaps even mocking). In terms of push-button malfunction, one "awfully mortified" user complained of a stuck push-button bell that would not stop ringing. She picked at it with her gloves and jabbed at it with her cane until an exasperated woman from the third story of the building poured a pitcher of water on her down below.[102] In another instance, hotel guests repeatedly heard bells ringing in a hotel although they did not push the button. Eventually, after hearing repeated rings, they began to tell ghost stories as the only plausible explanation. Only later, once an electrician came for inspection, did they learn that rats had eaten through the wires and caused them to form a circuit that "kept the bells ringing merrily."[103]

Accounts of user confusion often described button pushers as bumbling but harmless. A Texas newspaper, for example, reported such an event in 1899 when an "old gentleman" went to a firehouse to buy tickets for a firehouse ball and pressed a button in the station, unknowingly setting off a fire alarm. "The effect was electrical in every sense of the word," the paper noted. "From the air overhead—so, at least it seemed to the old gentleman in his bewilderment—men began to rain down."[104] Luckily, the article concluded, everyone enjoyed a good chuckle over the false alarm, and the innocent gentleman left with his ticket to the ball. Another story (1892) poked fun at a farmer unfamiliar with the ways that servants could trigger a front door of an apartment building to open upon the visitor's button push, noting that the farmer thought the wind had blown the door open.[105] Yet a third (1892) humorously recounted how a sleepwalking man showed up at one o'clock in the morning ringing a doorbell on Michigan Avenue in Chicago and demanding his pants, only to awaken from his unconscious state upon confrontation by a police officer.[106]

At the same time, people who lacked the skill or cultural knowledge to push a button and carry out a task incurred embarrassment for their technological illiteracy. The act of pushing a button could blur the lines among skill, status, and expertise, but it also could make these delineations more visible in certain contexts. Public shaming of those who made errors with buttons served a social function of separating out electricity users from nonusers; this created an "in" group of those with know-how about the machine age.[107] In some cases, reports of electric button mishaps simply noted incidences of electrocution or injury, but others of embarrassment and inexperience—button pushing gone wrong—also served as a newsworthy topic.[108]

These rejoinders against button pushers took a pejorative tact. Because push buttons often symbolized modernity and the strength of American technological achievement, those who could not use or understand buttons were marked as primitive and unworldly. For example, in a book recounting his experiences with Native Americans, author James Lee Humfreville (1897) took their unfamiliarity with buttons as a sign of their "savageness." He wrote to this effect about one particular encounter: "Jim was ill at ease among the surroundings of modern life. He did not understand electricity and its marvels. He could not comprehend how by pressing an electric button in his room a waiter forthwith appeared at his door."[109] This case and others proposed that every "civilized" and "modern" person should know how to push a button as part of a basic technological vocabulary, and that ignorance of this skill—or the appropriate social conventions for when to push—would give immediate evidence of the person's difference. With use of the button came a firm belief in American progress, racial/cultural superiority, and a staunch supremacy over electricity and its mechanisms.

By associating button pushing with a most basic form of every-day technical knowledge, more a life skill than a specialized skill meant for experts, newspapers and newsmagazines particularly differentiated between those "in the know" and those on the outside. Where electricians often disassociated themselves from buttons, which they perceived as common and ordinary devices, yet another level of hierarchy existed among laypersons by debasing those who couldn't even successfully press a button.[110] The act of pushing a button functioned as a societal litmus test, filtering out those too old or too foreign to participate in a growing machine culture. One had to prove her worthiness by demonstrating the basic technique of button pushing. Button usage exposed an embedded social hierarchy that delineated levels of expertise within layperson communities. Without the "right" use of the latest technology, one could easily find oneself portrayed as anachronistic or relegated to the punch line of a joke in a quickly evolving, machine-driven environment. A misstep with a button often produced slurs against the presser's race, gender, age, or class, and thus button pushing served as an outlet for expressing preexisting tensions and negotiating power relations. Although a change from pull to push meant that all users touched the same, in fact this technical change belied disparities in the rights, privileges, and skills of those who pushed buttons.

This problematic between the notion of accessibility—making buttons readily understandable and available for all—and access especially came to the fore in discussions about children as button pushers. Although educators espoused the importance of encouraging early encounters with buttons, accounts of everyday practices suggest that adults often viewed child button pushers—usually male youth—as nuisances, prone to using buttons in ways that violated social norms or irritated those around them. One of the chief complaints about children involved their penchant for ringing bells on a whim. For example, on the grandest scale—at the White House—author Mary Smith Lockwood noted the importance of button-triggered bells in an exposé on US presidents' wives that described a history of their household conveniences. Despite how far the presidential abode had come in its long-distance communication efforts, however, Lockwood discussed a comical scene in which President Harrison's young grandson summoned the whole of the White House with an innocent (or not-so-innocent) touch of a button, asking, "Did not little Benjamin, when alone one day in his grandfather's office, climb to his table, and by a touch here and there with his baby hand, set the whole force of secretaries, clerks and messengers on a chase to do his majesty's bidding?"[111] Although rhetoric of a perfect household controlled by push buttons circulated extensively in the late nineteenth century, an underlying concern remained about whom should have the right to press buttons and what kind of authority they could command; the image of a gaggle of staff members jumping at a button's demands (actuated by a small child) hinted at the potential backlash of a world populated by such devices. It also demonstrated that the value of reachability—putting buttons within reach of the average user (both physically and in terms of understanding) could come at a price.

Cases like this one with bells, although usually not on a nationally visible scale, led to technological interventions to manage child button pushers' behavior that was common and widespread. According to Rudolph M. Hunter, when granted a push-button bell patent, button pushers caused great irritation by "needlessly annoy[ing]" those individuals within earshot by pushing buttons for too long.[112] Similarly, inventor A. J. Oehring (1893) sought to remedy problems with button pushers who "meddled" and caused mischief, whereas J. C. McLaughlin (1889) constructed a new mechanism to withstand the misuses of fingers that pushed buttons "without any particular purpose" until they burned out and broke.[113] Indeed, an examination of patents reveals producers' many complaints about how children pushed buttons, and these rejoinders flew in the face of advertisers' romantic pronouncements about the reachability and universality of button pushing and button pushers.

Other measures were not so overt but nevertheless worked (and perhaps even more effectively so) to control how users pushed through invisible technical adjustments to buttons. In Rudolph M. Hunter's (1893) patent for a push button, for example, the document noted that "boys have a mania for pushing in and holding the button in closed circuit for the supposed enjoyment of hearing the bell ring or for mischief." He aimed to remedy this problem by creating a button with a bell that would automatically stop ringing after a set period of time, no matter how long the offending user pushed (see figure 2.4).[114]

Tips targeted at homeowners considering electrification were also sometimes advised to make buttons "secret" in their design and placement so as to prevent inappropriate uses by pranksters.[115] Similarly, railroad car designers responded to what they perceived as push-button abuses by making sure to locate push buttons higher than usual on walls to prevent children from using them.[116] Yet these interventions often generated rebukes as either ineffective or inconvenient. In the latter instance of railroad cars, this positioning generated backlash from irate adult customers who claimed that the "inaccessible location of the buttons," which caused the average patron to have to rise from her seat to press it, required immediate remedy.[117] Meanwhile, one observer perceived high positioning of buttons for children

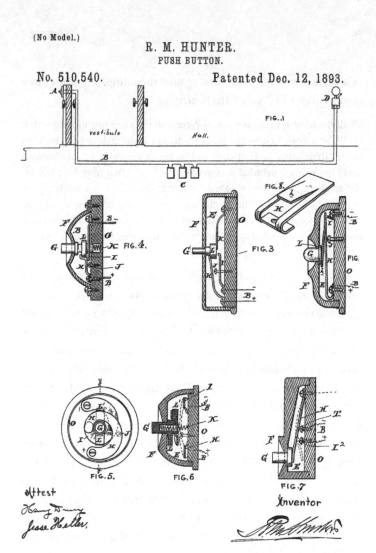

Figure 2.4
Patent for a push-button mechanism that would prevent boys from ringing bells for mischief.
Source: Rudolph M. Hunter, "Push-Button," Patent No. 510,540, patented December 12, 1893. Image courtesy of Google Patents.

as a useless technique to ward against meddling. According to a commentator in *Popular Mechanics*:

Of all the foolish notions recently recorded, the most unthinking is that of the New York architects who are putting the street door push buttons higher than usual, so that they will be above the reach of mischievous small boys. These deluded persons must be so old that they have forgotten all about kids. Otherwise they would realize that it is a mighty poor excuse of a boy that could be circumvented in any such trivial way. As it is a fair inference that the Manhattan boys are like boys everywhere else, a high door bell will present merely an added temptation to their enterprise.[118]

This strategy of putting buttons out of reach failed to take into account the enterprising ways of youth.

It was not as though push buttons stimulated youthful pranks where none had existed before. In fact, some lamented an influx of push buttons as signaling the end of youthful amusements rather than stimulating new ones. According to a Mr. Stoggleton interviewed in the *Boston Daily Globe*, fun could once be had for the small boy with the door knocker "whose thunder reverberated through the hall and filled the house" and the old-fashioned bell pull by "yank[ing] the bell-pull out to the limit, causing the bell to fly almost off the spring." In the present moment, however, he suggested that "there is no such fun in pushing in a push button. You can press that in perhaps a quarter of an inch, and that is all you can do with it. ... It is dry fun," and he predicted that ringing the bell and running as a prank was quickly disappearing as a childhood pastime.[119] Much as some mourned the loss of individuality of touch that door-knockers and bell pulls provided, this observer proposed that the simplistic and less dramatic push button symbolized an end to childish fun. Evidence indicates, however, that buttons continued to attract youthful fingers.

A problematic conflict existed, then, between desires to educate children and make them "good" consumers of push-button
electricity and their tendencies toward pushing in ways deemed
out-of-control and for the mere pleasure of pushing. Blurry
lines between push-button-as-play and push-button-as-control-
mechanism meant that bells were vulnerable to the hands that
found them worthy of exploration, and they caused an affront
to the ears when one did not expect or want to hear them. Buttons seemed to operate too well, as a child's touch could turn
the mechanism into a liability rather than a convenience. These
kinds of problems with push-button bells demonstrated the
complexity of digital command.

3 Servants out of Sight

A number of experiments were shown, illustrating both permanent and induced magnetism; and a variety of forms of electric bell were exhibited, from very sweet-toned churchbells of a small size to a contrivance intended to wake up servants, which, when once the button is pushed, goes on ringing incessantly till the tormented servant goes to the bell and pulls a handle.[1]

Bells often fell victim to the hands of pranksters and mischief-making youth because the practice of ringing by push button to summon service involved not only reducing hand effort but also creating ergonomic environments where a button always existed within easy reach of the hands that desired them. In fact, the more servants, call boys, bellhops, and other service workers were relegated to remote parts of the buildings they inhabited as part of a widespread social shift of physically separating employers from employees, the more architects, electricians, and advertisers advocated for bringing push-button communication into the personal space of those summoning their presence.[2] Push-button communication served as one strategy to make servants—and their work—"invisible"—to those who commanded them.[3] By relegating servants to certain parts of the home, homeowners could achieve both a strategic and symbolic distance, while a push-button signal could command the servant to appear at will.[4] To this end, sophisticated designs cropped up for homeowners to disperse push buttons throughout their property—a house outfitted with such devices might have buttons in the floor, under the dining room table, in the hostess's lap, or on a bedpost (see figure 3.1).[5] These uses suggested that buttons might function as extensions of the presser's body, a kind of prosthetic mode of control that could make communication seamless while carrying out other tasks.[6] The basic premise underlying the arrangements of push buttons in domestic and commercial spaces involved encouraging electricity users to exert control from a single position rather than move around. Appeals took different approaches, but in general they advocated that push-button control could benefit users who couldn't move around to communicate (as in the case of invalids and hospital patients) and who needn't (those whom demanded luxury or whom could take advantage of positions of power). By making push-button control diffuse, always within reach, and gentle to the touch, designers and users together created a vision of buttons as effortless and discreet, yet powerful. When physical surfaces could transform into buttons, even a table leg or table could conjure up electrical effects.[7]

These embedded electric buttons offered—but did not guarantee—the promise of efficient, effortless communication; a homeowner might lift a finger (or sometimes a foot) to carry out her wishes but no more. Advertisements for such devices emphasized that one could make a call "without stirring" and with "but a very slight pressure for the finger."[8] So as to minimize bodily effort, inventors created buttons like the "presselle," which one could transfer from any side of a table or attach it by a hook to one's clothing—to avoid "fishing around with the

Table Clamp

For use on dining tables, card tables, chairs, desks, and other movable furniture.

The neatest table push on the market.

Clamp and Push......40c

Silk Cord, double conductor, per yard....15c

Cut ½ size.

Figure 3.1
Buttons clamped to tables or hidden in floorboards or bedposts were meant to make homeowners' control of household affairs diffuse and unseen.
Source: J. H. Bunnell & Co., *Telephone Catalogue and Manual of Telegraphy with Description of Instruments Adapted for Use on Private Telegraph Lines*, 1899. Image courtesy of Collections of the Bakken Museum.

foot."[9] Some designers went as far as to patent novel arrangements for embedding pushes in desks that could slide in and out so that "the push buttons are always within convenient reach even though the top of the desk is full," a common arrangement for push-button managers (see figure 3.2).[10] In visions of the push button, electrified home, any surface could function effectively as a site for push buttons as long as a finger could easily reach the button.

In the case of housewives, especially, advertisers promoted a relief from housework through digital command, whereby

Figure 3.2
A typical depiction of an armchair manager who pushed buttons to summon his employees.
Source: "Push Buttons," *Illustrated Catalogue and Price List of Telegraphic, Electrical & Telephone Supplies* (January 1888): 126. Image courtesy of Collections of the Bakken Museum.

the "frailer sex" could enact control from a sedentary position with a single finger.[11] Author Francis Cruger Moore (1897) advocated for control with a push in a treatise on "how to build a home," emphasizing that buttons should save the mistress from "an uncomfortable walk across the room."[12] In addition to encouraging housewives to control bells or other electrical appliances without moving about, arguments made for push-button installations often centered on ways to minimize the lady of the house's efforts to communicate with servants so they could carry out household labor instead. For example, upon rising in the morning, a woman could "discourse with the cook" about breakfast without having to put on clothes and descend the stairs.[13] The *Sun* paper of New York caught wind of this phenomenon, noting that, "it is now impossible to see how the lady of the house communicates with the kitchen while a meal is in progress." The author marveled that "the call button has become a mere electric button on one of the table legs."[14] Manufacturers and advertisers typically romanticized push buttons and their users, depicting fingers as uncomplicated, unfettered modalities and electricity as safe, touchable, and ready to conjure one's wishes into being. Stories in magazines such as *Country Life* liked to perpetuate these myths, marveling that a mistress could go as far as to hide her push buttons inside a sofa, thereby making visitors unaware of her electrical control so that "the bringing in of tea or the appearance of the butler with hats and wraps was spontaneous."[15] Electricians disseminated this view to make an electrical household more attractive, as if push buttons could conjure servants from thin air. The housewife could sit in her dining room, for example, and instead of interrupting a good story at the table with a noisy hand bell, she would noiselessly push a button with her foot to "summon the servant to do her bidding, and conduct the affairs of the table quietly, easily, and magically."[16] A "mere touch" of this kind purported to make the act of summoning an effortless one. These inventions and idealized scenarios reflected patterns of change occurring at this time period that aimed to mechanize household work and scientifically manage the tasks of everyday life through a movement often referred to as "domestic engineering."[17]

Comments about push-button communication explicitly noted the supposed benefits of keeping "the help" away from those pushing buttons. A sick person, for example, could make requests via a button attached to a cord at the bedside or fastened above a pillow, thereby "summoning an attendant whenever required, and saving him their presence when he would be alone."[18] Similarly, a strategically placed button could enable a woman in her home to avoid all contact with her hired help in the course of a day. An article about such an invention remarked, "By pressing an electric button a large sign appears in the kitchen indicator, saying, 'You are discharged.' In this way the mistress of a home can dismiss her cook and stay quietly locked in her bedroom until that lady has departed."[19] The motivations behind these devices suggested that the button pusher could "toggle" the servant's presence "on" and "off" at will so as to minimize any "unnecessary" interactions. Thus, positive characterizations of push-button communication between housewife and servant, often espoused by electricians and domestic engineering experts, imagined an environment where servants functioned in the following manner: "Little electric buttons, to call at will any one of the different servants, are put in almost every room. Touch a button of one color, and the butler appears; of another color, and the maid, and so on."[20]

Importantly, the acts of pushing a bell to call a servant, ring up an employee, instruct a chauffeur, or get a bellhop's attention inevitably involved power dynamics and the expectation that the button pusher could instantaneously summon anyone or anything she wanted; button pushers occupied positions of privilege. Yet those pushing buttons did not always reap the benefits of this privilege. Miscommunications and user errors often occurred. Servants ignored calls and refused to respond in a timely manner or at all. The success of pushing buttons depended not only on properly working mechanisms but also on a docile and willing workforce, both human and technical, that would pop up at a moment's notice ready to do the pusher's bidding. As discussed previously, digital commanders used buttons for all manner of summons, commonly pushing to trigger the electric bell that would announce visitors' arrival to one's front door, the call button that would beckon servants from a distance, and the annunciator that would signal hotel staff of a guest's wants. Push buttons were meant to mitigate labor typically associated with communication and reduce it to a mere finger press, but under many circumstances distance seemed ever-present and even insurmountable. Button pushers often lodged complaints about the nonverbal, one-way communication that buttons facilitated, particularly with regard to the bells they rang, just as those waiting on the other side of the button expressed dismay about their treatment.

Usability issues manifested when patrons used buttons improperly or violated the implicit rules of etiquette between staff and guests. One of Theodore Roosevelt's Rough Riders caused such a ruckus at a New York hotel by using an electric button as "target practice" with his pistol. When asked why he shot at the button by the clerks who rushed to his room, he replied that he was "thirsty, too tired to get up, and was trying to ring the bell."[21] *The Columbus Dispatch* also reported on a kind of push-button "abuse," whereby a young boy (left unattended by his father) summoned a bellboy from his hotel room, demanding that the employee send "some one I can say my prayers to, and send him quick" so the boy could go to sleep. According to the witness, the push button "must have been worked to the limit," given the way the bellboy had rushed to his aid, assuming the frequency of button pressing was meant to denote an urgent communiqué.[22] The article admonished this overworked use of the button—and the unsupervised button pusher—for taking advantage of a service. When attendants perceived that people pushed buttons too much or inappropriately, they reinterpreted button pushing not as a courteous service but rather as a forceful, frustrating, and excessive demand for presence.

Indeed, complaints often arose when one party felt that the push or its response did not follow explicit or implicit social norms. The experience of George Bidwell (1890), recounting a frustrating interaction at a hotel in his autobiography, is a telling one: "When I was ready to go out I had occasion to call a servant, and touched the electric button," Bidwell wrote.[23] Upon getting no response, he tried again: "Soon I touched and held down the button for a longer time, and again waited in vain. In my then nervous condition I lost both patience and temper, and continued the pressure on the button." Eventually, he described how a gaggle of servants came running in response to his "concert of at least a hundred bells going." Bidwell concluded, relieved, that servants answered his calls promptly for the duration of his stay, and in hindsight he found the experience "humorous" enough that Mark Twain could have written a whole chapter on the subject. The autobiographer had come to expect instantaneous servitude whenever he pushed a button. Bidwell did not find his entreaties for service as forceful or unpleasant; he expected buttons—and the people made to heed them—to wait at his beck and call.

Indeed, push-button magic only worked when attendants seemed to appear instantaneously at a moment's push, as though they waited in the wings for a cue. Hotels tried to produce this kind of luxury service by installing push buttons in dining rooms. Cooks could communicate to wait staff when they had prepared food for customers by triggering a flashing light that corresponded with the appropriate table number.[24] Dining patrons did not always find these interventions helpful or indicative of faster service, as one reporter surmised: "Although the electric bell has invaded the hotels in the interior, its use is not yet allowed to disturb the leisurely habits of the waiters." The observer humorously noted a sign above a push button that read, "Push in the knob. If you do not get an answer in fifteen or twenty minutes, push it again."[25] This lag time between push and response seemed to break a contract inherent to the call-and-response system.

At the same time, guests listed their own grievances at feeling put-upon by bell ringing. In the case of call bells, hotel staff could provide wake-up services to guests by pressing a button in their office that would trigger a bell in the guest's room. Although this seemed like a luxury service for patrons, Morris Phillips (1891) warned tourists in his practical travel guide that "there is no escape from [the bell]; you must get out of bed to stop the ringing."[26] This "convenience" proved merely irksome to some, even inspiring "profane language and premeditated murder on the part of the victim," who would hear "a cross between a cow bell and a Wagnerian overture."[27] From the staff's perspective, the simple push of a button saved a trip up multiple flights of stairs and avoided disturbances of other guests, when the choreographed interaction between button pusher and staff went according to plan. For guests, however, this form of communication might do more harm than good by offending their aesthetic sensibilities and forcing them to exert physical effort through rapidly pushing the button to make the sound go away. These negotiations demonstrated that push-button configurations were meant to reduce effort and labor—but only for some. Whereas a hotel patron might sit comfortably in his room, or a housewife might sit reclined at her dinner table, noiselessly and politely choreographing all the elements of her meal, these minor efforts could only transpire with someone else's effort hidden from sight and acknowledgment. It is no surprise, then, that the digital commander who communicated and commanded from a distance often incurred criticism for laziness and abuse of power (see figure 3.3).[28] Indeed, one writer used electric buttons as a representative example to illustrate that "luxury begets ennui," whereas "hard work leads to happiness."[29]

Many of the problems involving electric bells for communication practices arose from broader difficulties related to domestic service in the late nineteenth century. A high degree of turnover among servants at this time period (with most not staying longer than twelve months) meant that systems were necessary to understand how to maintain—and even use—push-button bells. Push buttons could only improve the quality of home life if all of the individuals involved in the communicative act participated in its construction and agreed on the meaning of the signals being given.[30,31] To this end, Isabella Mary Beeton (1890), in her guide on how to decorate and manage one's home, advised that each push button should have a label so that its user could easily identify the room with which it corresponded.[32] Despite such systems, homeowners in large households quickly learned that "unruly" servants would find easy excuses in the temperamental equipment for their insubordination. Servants could resist to some degree, for example, by blaming malfunctioning bells; the button could only discipline the ear that heard it or the eye that saw it pressed. Servants would give an excuse "from time to time to cover their negligence that 'the bell didn't ring,'" or they would "decline to answer [bells] as long as possible" until they eventually tended to the call and "protrud[ed] [with] an

A LUXURIOUS FELLOW.

MESSENGER-BOY. — Call, sir?

CADSEY SCADDS. — Ya-ah. Just step across the room
there, and touch the electric button. I want me valet.

Figure 3.3

A satirical cartoon played on a common assumption of button
pushers' laziness, depicting in the extreme how a man called for his
messenger boy for the sole purpose of pushing the call button to call
someone else.

Source: "A Luxurious Fellow," *Puck* 29, no. 732 (March 18, 1891): 51.
Image courtesy of University of Michigan Library via Google Books.

irritated, murky face," in the words of one dissatisfied button pusher.[33,34] Buttons did not serve up magic at all if the hands that pushed them garnered no response or produced a begrudging attendant.

These tensions between employers and employees and patrons and staff extended beyond domestic contexts, too. For example, travelers had long abhorred the lack of communication on streetcars between passenger and conductor without resorting to "frantic endeavors."[35] One complainer (1889) offered that to solve this problem, "some simple device in the nature of the familiar electric push-button may supply this want where electricity is used as the motive power."[36] Indeed, in the years immediately following this comment, streetcar companies did experiment with push-button bells in cars, but with varying degrees of success.[37] In an 1895 letter to the editor titled "About Those Buttons: One Man Who Saw How They Didn't Work Very Well," a user described his initial enthusiasm about the new communication technology, which might prevent him from having to wrench his neck or risk falling in attempts to get the conductor's attention. Alas, he described a series of failed attempts at communication. In one case, he pressed the "cute little black button" only to receive angry admonishment from the conductor that the button wasn't working. In another case, he witnessed a woman hold up a "beautiful little child" to press the button when her stop arrived, only to be shouted at by the conductor: "We don't stop when brats fool with the buttons."[38] These instances suggested that technical failures and misuses of the technology had made conductors suspicious of the electric button, so easily pressed and therefore abused, leading them to thwart passengers' efforts. When city commissions took pains to make push buttons mandatory on cars to ensure riders' safety,

conductors spoke to newspaper reporters to argue that the interfaces "proved a useless accessory," and they noted that passengers did not use them. President J. M. Roach of the Union
Traction company (1900), a popular streetcar manufacturer in
Chicago, argued that buttons were perceived "as a joke by the
traveling public, and would cause much annoyance were they
in general use." Instead, he suggested, signals should remain
under the control and purview of trained employees.[39] Conflicts
over proper channels of communication, authority, and power
made the button a contested object in streetcars, with each kind
of user pinning blame on the other. Whether users did indeed
pervert buttons' intended purpose or conductors were particularly sensitized to a loss of control they once possessed remains
unclear.

In other kinds of workplaces, similar confrontations around
push-button communication occurred. To meet the goal of
achieving maximum efficiency in many contexts, organizations
installed a growing contingent of managers to reinforce the values set forth by the scientific engineering movement. Typically,
historians have described this move as a rise in the employment
of "'head' workers"—higher-ranking employees whose tasks
shifted from manual to mental labor.[40] Yet this often-advanced
perspective maintains a traditional dualism, wherein brain and
body occupy separate realms, and it also overlooks a more tren-
chant divide between workers' hand practices. Those sitting on
high in offices still used their hands, but their employees complained about how they used their hands in the form of digits
that directed other hands to perform work. Individuals working
"in the trenches" to carry out dirty, dangerous, and physically
exhausting occupations often redefined button-pushing hands
as "nonlaboring" and button pushing itself as "nonwork." These

labels reflected angst about new kinds of work activities in office settings that prioritized sedentary labor and dispensing orders through hand gestures and bell ringing.

Desk workers began managing their employees through these remote hand practices in part due to growing physical separation between types of employees, as managers increasingly directed those doing grueling or menial work from a distance. Push buttons, as communication technologies meant to accommodate this distance or to make it possible, therefore often exemplified a set of priorities for separation, stratification, and remoteness of employees. These reallocations of space, which created front stage and backstage conditions, also involved rearranging hands and calling on them to perform activities to cope with distance using their fingers.[41] With workers' hands made invisible from their bosses and vice versa, tensions about manual effort and fears over eroding "human touch" both intensified.

Particularly in the case of workplace communication, button pushing often drew harsh commentary for its comical claims toward "efficiency," which promoted unequal hand practices. Managers often used push buttons to transmit orders to or summon lower level employees; although the push of a button to call at a distance could streamline these interactions, complaints routinely cropped up about buttons in reference to accusations of armchair managers' "abuse" of rank-and-file employees. Detractors worried that push-button communication created a kind of impersonal, impermeable distance between managers and their staff. Although laborers never united formally in their efforts to protest button pushing as a workplace practice, anti-button sentiments often appeared in publications ranging from trade magazines to newspapers, citing the fact that the "push-button-and-messenger system of communication between the different parts of the office has a strong demoralizing effect."[42] Complaints labeled slothful, entitled men in high-ranking positions who often rose to prominence in the scientific management era as the culprits.

A common concern stemmed from the ways that push-button managers tried to limit physical contact with those of lesser station. For example, to gain access to the mayor in Boston, an 1895 article in the *Boston Daily Globe* explained that, "the electric button is king at city hall."[43] Indeed, the mayor sat protected in his office behind a closed door, and a messenger sat out front with access to a series of push buttons. His touch sent a signal that determined who could enter, and the combination of pushes changed each week to prevent unauthorized persons from entering. Acting as a gatekeeper, the button, the paper reported, "quietly turns down a good many good men."[44] By controlling the comings and goings of visitors to city hall, buttons acted as disciplinary devices—reinforcing the boundaries of the mayor's office space so that his attendant need not lift more than a finger. A single push could execute orders without uttering a single word. Resentment often cropped up due to these kinds of practices.

Buttons' design and one-way communication style meant that users could exchange limited (if any) information via ringing for one another, and thus pushing a button rarely conveyed intent or the "personal touch" of the operator. Employees called into their bosses' offices rarely knew why they received a call, and thus they might arrive already sensitized to the fact that the button seemed to beckon them without apparent cause.[45] They might also grow fatigued from constant demands for their physical presence, enduring the indignation of digital calls that never worked their pushers into a sweat. In this regard, responders to call buttons often interpreted calls from a distance as derogatory and physically taxing. These feelings intensified stratification between managers, perceived as digital commanders who gave directions with their fingers, and traditional "manual" workers beholden to their calls. When digital commanders discussed the benefits of push-button systems, they focused on less energy expenditure for themselves, as opposed to those summoned by the push: "It is surprising how many steps the use of this little contrivance saves, and how greatly it facilitates the transaction of our office business," noted one writer whose office began employing such a system. He continued, "The [push button] saves all such troubles, and enables the manager, without leaving his seat, to communicate instantly with all the principal persons employed in the concern."[46] Communication in these contexts was focused on having employees work on demand or be at the manager's beck and call, and various features were employed to achieve this goal.

Elaborate systems to facilitate these calling relationships made "pages" or "call boys" wait perpetually in a state of readiness. In political arenas like the state house in Boston and the Congress building in Washington, DC, push-button arrangements helped political officials to request documents and other services from their pages without rising from their seats. In the former case, 240 desks were outfitted with buttons wired to two indicators at the rear of the chamber; an annunciator would indicate the row and seat number of the person pressing the button to alert a page's attention (see figure 3.4).[47] This device and others like it were touted for reducing the frequency or effort of hand gestures required for attracting someone's attention.

In Congress, the act of pushing a button eliminated the "pernicious habit" of clapping one's hands to get a page's attention;

Lamp Annunciator Set in Pages' Table.
Note Flexible Cable From Wall Outlet
Which is Provided With Separable Con-
nectors

Figure 3.4
An illustration of an annunciator system embedded in a desk, which
functioned by lighting up to attract a page's attention.
Source: "Push Button and Annunciator System," *National Electrologist* 29,
no. 6 (1922): 52. Image courtesy of Ohio State University Library via
Google Books.

at the state house in Iowa, representatives no longer had to snap their fingers.[48] Of the latter case, one observer noted in 1884, "To those who have seen, in the midst of an exciting debate, members violently pounding their desks and snapping their fingers in fruitless endeavors to secure the services of a page, the advantages of an electric system of calls, whereby the noiseless pressure of a push button secures the prompt attention of the desired messenger, will be very apparent."[49] Where this view optimistically imagined a quiet touch replacing a noisy entreaty, much as the quiet bell replaced the noisy door knocker, those being called interpreted such a change as significantly detrimental to their perceived role and worth:

The page boys have registered a strong kick against the electrical apparatus need in the chamber of the House of Representatives. Heretofore they were all allowed to sit on the steps in front of the Speaker's chair and wait upon members, who indicated by the snapping of fingers or clapping of hands that a page was wanted, and one of them would respond. In these times the boys were associated directly with the members, heard everything, and were many of them statesmen in a small way. "Well," as one of them put it, "we're not as good as hotel bell boys. We are huddled up in this room all day, see nothing and hear nothing except when on an errand, and that is all due to the darn electric annunciator they got in here." ... Now when a member wants a page he simply presses a button at his desk.[50]

Of note, page boys lamented that the new method of calling shifted their physical presence out of sight, thereby revoking their rights as "statesmen" who could see and hear political transactions. As in the case of housewives summoning servants, pages did not wish to be put out of sight until needed. Whereas electricians hoped that a push-button summons via a finger push would make pages physically more efficient, those at the receiving end did not take kindly to requests that sought to toggle back and forth between presence and absence, interpreting the push as injurious to their participation in the political process.[51] These calls, used in political offices as well as factories, shop floors, and hotels, consistently worked to keep young employees engaged and accountable.[52] Such assaults on call boys' attention generated the impression among many that "at no moment of the day is a man safe from the summons of the push button."[53]

Some experiments took this commonplace system one step further to make the page's presence increasingly malleable. In Congress, an electrochemical annunciator allowed members to calibrate the degree of urgency for their requests through touch. According to the *Washington Post*, "A Representative wishing to send a page on a trifling errand, lightly touches the button on his desk. Instantly his disk in the case turns a pale brown. The boy on guard notices the change and saunters off to answer the call." The article continued, "Another member is in a hurry for a certain document, and gives his button a steady push. The little metal button at the other end of the circuit gets brown, then black, and as the pressure remains it turns red. This is a signal that the member is in a hurry, and off rushes a page."[54] By tying a sense of timing to the touch, the system aimed to overcome some of the limitations of nonverbal, one-way communication across distance. Yet when taken to the extreme, the touch might communicate too much:

When an impatient, hot-headed member, who has had a bad night, wants a page, the antics of the disk at the other end of his line are wonderful to behold. The little object gets brown, black and green in short order, and then turns a livid red. There is no let-up in pressure by the member, and the indicator fairly outdoes itself in trying to reflect the feelings of the irate Congressman. It literally sweats blood, for a crimson stream trickles from the disk and spreads over an area of an inch or more. When the boy arrives at the scene of the disturbance, and the member's thumb is lifted, the red spot gradually thins out and disappears.[55]

This transmission of affect across distance imbued the button pusher's touch with markers of intensity that went beyond a ringing bell by wedding emotion and tactility. By providing a visible sign of finger pressure, the push-button device made the hand all the more demanding. The call button that changed colors demonstrated how the digital commander could transmit a forceful message even without a great deal of finger force.

Although employees might have limited opportunities for resistance, like servants they found ways to turn the button-pushing act on its head. For instance, they often called to attention to—and derided—their bosses' frequent incompetence with buttons. Stories circulated to make these ridiculed button pushers amusing fodder among the ranks of manual workers, as in the case of an army major who unknowingly sat on a panel of buttons on his desk, repeatedly calling members of his company without any awareness he was doing so. The narrative retold how, after a number of frustrating encounters in which "confusion reigned supreme," the major "vacate[d] his seat precipitately," becoming the butt (both literally and figuratively) of the joke.[56] In another instance, a man meeting with "prominent officials" accidentally pressed a button on his desk that was intended for removing unwanted visitors from his office. This errant push led three officers to "rush into the room, all heavily armed" and caused the visitors to leave without knowing the reason for their eviction.[57] The incident exposed the inadvertent button pusher to humiliation for his poor handling of the situation. Readers could have an easy laugh at the expense of the blunderers while mocking the inefficiencies of using buttons for communication and control. By poking fun at managers, leaders, and other wealthy men, manual laborers could reframe digital command as silly compared with those who truly "worked" with their hands.[58]

Given the many ways that manual workers could circumvent or impede the calls that demanded their presence, designs appeared to manage "servant problems" beyond the purview of the original push button to increase accountability.[59] One inventor designed a button with a magnet that would indicate whether a bell had rung. Another suggested that a homeowner should have a button installed in her bedroom—which attached to a loud and continuously ringing bell in the servant's room in the attic—so that requests would not go unheard.[60] Others still, such as a push and bell system used in a hospital, rigged a patient signaling device that would continue to ring until the nurse came to the patient's room to deactivate it.[61] These buttons worked as disciplinary devices as much as communication devices by attempting to enforce docile behavior, provide oversight of staff, and segregate help from the spaces that homeowners embodied.[62] Generally speaking, electrical experts recommended "energetic" or "aggressive" bells for large spaces where a servant would need to respond from a long distance.[63] Within the context of broader problems related to domestic labor and laborers, these inventors crafted solutions that put control firmly in the hands of button pushers to manage household staff and the communication that occurred between them.[64]

Buttons could "go a long way toward helping [the housewife] solve the vexing servant problem," some believed.[65] However, they only functioned "properly" when "invisible" hired help responded to the button pusher's beck and call; a servant who refused to heed the message or took a long time to do so violated the contract of instantaneous service. Indeed, in a book titled *Home Mechanic*, author John Wright (1905) noted, "When a bell is rung it should be answered *immediately*, and the arrangement should be such that there can be no excuse for delay" (italics original).[66] This "on-demand" approach to servants' work meant that help must remain in a state of readiness at all times to create a seamless transition between button push and effect. Such transactions between pusher and responder suggest the underlying motives of button pushing as part of a larger effort to compel other people's bodies to work toward the aims of those who directed them. Pushing buttons acted as one strategy for encouraging a docile workforce that would remain out of view until needed and then conjured seemingly from thin air. Although pushing electric buttons might do away with noisy calls or strenuous hand actions, these acts continued to receive a negative response for the ways that they demanded a rapid response and a willingness for hired help to shift between presence and absence without ambiguity. Friction often surfaced between the "pusher" and the "pushed."

II Automagically

4 Distant Effects

The rushing train, entrusted with a thousand lives, is checked by the motion of a single arm. The complicated machinery that whirls, and groans, and labors through the town-like factory, may be set in motion or arrested by a single touch.[1]

"Suppose you could push a button and—thus obtain your dearest wish—love, fame, wealth, or power—and at the same time cause the death of an unknown person in China? Would you push the button?"[2]

Where the practice of pushing a button to ring a bell worked to govern household and workplace help at a remove, other push-button acts functioned to demonstrate how one could command electricity and achieve spectacular effects across distances. Unlike servants or pages, who could refuse to present themselves in a timely manner (or at all), electricity—when it functioned properly—could appear instantly. By concealing electrical wires behind walls, wallpaper, and push-button faceplates, and by putting buttons in separate locations from that which they controlled, the severance of cause and effect could enable a thrill for spectators. One event that demonstrated this promise of distant effects and digital command involved a choreographed display by William Joseph Hammer, an associate and employee of

Thomas Edison, who staged a reunion for his classmates in 1885 on New Year's Eve. *The Electrician* journal reported on the event to electricity enthusiasts:

The members of the class came from all parts of the country to revive the memories of the past, and to witness some novel experiments in electricity. Those who rang the doorbell were anxious to let go as soon as possible, as some person had attached an electric wire to the knob. On advancing to shake old classmates by the hand, the new comer accidentally stepped on an electric button, and every light in the house was extinguished.[3]

Not only did the unwitting visitor stumble through the electrical house experiencing shocks and thrills of a totally foreign nature, but the scene was made all the more impressive by Hammer's omnipresent control. According to an official booklet published to record the momentous occasion, "The innumerable electrical devices shown during the progress of the dinner were all operated by Mr. Hammer, who controlled various switches fastened to the underside of the table and attached to a switchboard, which rested on his lap, while the two cannons were fired by lever switches on the floor, which he operated by the pressure of the foot."[4] Although the various switches, boards, and buttons employed by the host would likely have incurred notice by participants, Hammer made efforts to conceal these control mechanisms underneath his table, on his lap, and beneath his foot. This configuration gave him the role of puppet master pulling invisible strings, making electrical curiosities spring forth seemingly from nowhere with his body wholly wired into the system.

Merging concealment with electricity and control, the first electrical house demonstrated with spectacular versatility an important aspect of digital command. It identified a desire to focus on reachability as a key component of electrical control, and it involved the notion that controls should exist within one's personal space with no need to reach for them. In Hammer's experiment, push buttons functioned as a kind of "prosthetic" that enabled him to conjure swift effects with a simple touch of a finger.[5] At the same time, the push-button spectacle demonstrated that buttons worked best when they could create magic. As F. J. Masten (1893) observed, "men like to push the button and let others do the rest, but we must not forget that these other fellows do not allow patrons to go behind the scenes; there must be no investigation of the secret springs and causes of action, for mystery is just what the button pushers like."[6] In significant contrast to the ways that housewives or managers commanded their employees—often without garnering a quick or pleasant response—electricity could produce the instantaneous magical effect so desired.

Although Hammer's demonstration might have thrilled friends (other electrical experts), proponents of push-button control envisioned digital command as a broader practice for the masses by stressing that buttons should be "within reach of children and the domestics" and "pliable to man's will," with the proposition that anyone could take on the role of a digital commander.[7] As with educators, the electrical industry also routinely linked mere touch particularly with youth and femininity; the act of touch signified humans' ability to "tame" and master electricity, to turn any body from a subject to a master. To this end, push-button boosterism often centered on staging idealized spectacles of young girls pushing buttons to demonstrate how digital command could apply to the (supposed) lowest common denominator. One of the most widely discussed events took place on October 10, 1885, at 11:10 a.m., when 11-year-old Mary Newton pressed a telegraph key (whose bulbous end was commonly referred to as a "button") to detonate dynamite that blew up a mine 1,000 feet away, spewing bits of rock into the air and making the earth tremble.[8] Onlookers described the girl as the picture of poise and femininity, possessing a delicate touch in the midst of such power (see figure 4.1).

Common references to the button-pushing girl emphasized the violent, unexpected, or profound effects that her hand could achieve. Physician B. E. Dawson registered his wonder that "the little child's delicate finger may touch a key and blow up Hell Gate," referring to Newton's push-button destruction of a bridge in New York.[9] Similarly, author Elisha Gray marveled that "the most delicate touch of a child's finger will be sufficient to release enough energy to destroy—in the twinkling of an eye—the largest battle-ship that ever plowed the ocean."[10] In another instance, according to Dr. R. A. Torrey in an allegorical Sunday school lesson to a group of children about Mary Newton's push-button act, "I said 'there is but little strength in my finger, but when I pray I put forth my weak finger and touch the arm of God, the arm that moves the universe, and that mighty arm moves to do its work.'"[11] This imagery of a weak finger activating a "mighty arm" served as a fitting representation for how push buttons could offload human labor onto machines by enabling a light touch. Each of these descriptions recalled the image of an unbalanced scale, dating back to the tale of David and Goliath, in which "tiny efforts balance out mighty weights." Such a feat had long seduced with its promise of making the weak strong through mechanical interventions.[12] If hands no longer strained in effort, then humans could proclaim they had truly achieved a "reversal of forces."[13]

NEW YORK.— GENERAL NEWTON'S LITTLE DAUGHTER TOUCHING THE ELECTRIC KEY
WHICH EXPLODED THE DYNAMITE MINE UNDER FLOOD ROCK.— SEE PAGE 135.

Figure 4.1
A young girl blew up a mine by actuating a key—often called a
"button"—to demonstrate the marriage of delicate femininity with
technical prowess.
Source: *Frank Leslie's Illustrated Newspaper*, October 17, 1885: 144. Image
courtesy of Internet Archive.

Experimentation with distant control at a touch, such as Newton's mine explosion, began with telegraphy. Not only could one send a message to someone at a location removed from the sender, but telegraphic communication also seemed to challenge how message senders and receivers thought about their bodies, for "Telegraph lines ... appeared to carry the animating 'spark' of consciousness itself beyond the confines of the physical body."[14] This notion of extending one's consciousness through wires took on a particularly compelling character at world's fair events, when a telegraph wired from a remote location to machinery at the event site could activate steam engines, lights, and other displays on the fairgrounds, as though an invisible hand had reached across thousands of miles to push a button. Such long-distance button events became increasingly popular—and even standard—at the turn of the twentieth century. These spectacles featured prominent figures and politicians, especially US presidents and their family members, as the button pushers to emphasize the political and physical importance of touch across distance.

It is no surprise that presidents and others would have employed the telegraph to appeal to fair audiences from a distance. This mode of communication and control greatly reduced expense and time commitment, and it also reinforced the ideals of the world's fair and its rhetorical promise—as an "artificial realm" outside everyday reality—to expose visitors to the most cutting-edge technology in a celebration of national achievement.[15] Visitors to a nineteenth-century fair would typically have far more electrical encounters than they had in their lives up to that point, experiencing novelties such as moving sidewalks, Ferris wheels, electric light displays, and animated fountains.[16] By disguising the dangerous and unsavory aspects of these devices and cloaking them in spectacular and dramatic demonstrations, fairs served as powerful tools of persuasion through community engagement; these events were often designed to "win the hearts and minds" of participants, advancing hegemonic and imperialistic policies through expressions of regional and national pride.[17] Never neutral in their presentations or representations, expositions in their other-worldliness encapsulated technological fantasies and fears, often acting as a suture between everyday experiences and larger societal concerns based on race, class, and gender.[18] Through press coverage of button-pressing activities, because their remote nature meant that few experienced the event in person, lay audiences were asked to consider how push-button interfaces could both sensationalize electricity and profoundly extend their sense of touch. Media discourses provide insight into the ways in which journalists framed issues of human–machine relationships in the context of national pride, spectacle, and performativity with technology.

Descriptions of male button-pushing hands in these contexts—US presidents and other males holding high-level political offices—referred at once to masculine virility, natural disaster, and masterful taming of electricity.[19] These touches were portrayed as summoning the mightiest power, commanding electrical forces to do the pusher's bidding (see figure 4.2). Reports often used language such as, "The electric current will come bounding fresh from the hand of President Cleveland."[20] Language of this kind emphasized the generative power of hands, or as science author Arthur E. Kennelly (1891) put it, "The relation between electricity and vitality may be so close as to amount to identity."[21] In broader discussions of electricity, individuals vacillated between viewing nature as both "an object of conquest" and "an ally with whom mankind was in

PRESIDENT CLEVELAND TOUCHING THE "VICTOR" TELEGRAPH KEY

Figure 4.2
President Cleveland touches a telegraph key, commonly called a "button," to start machines into action at a distance.
Source: "The 'Victor' Key Opens the World's Fair," *Electrical Review* 22, no. 12 (1893): 157. Image courtesy of University of Michigan Library via Google Books.

direct dialogue."[22] As the fleshy body of a presidential figure connected with an electrically charged telegraph key ("button"), both human and machine would animate beyond the capacity of either entity alone.

Once again referring to this rhetoric of masculinity, in the words of a self-described "intelligent Englishman" who attended the World's Columbian Exposition of 1893, President Cleveland "showed himself in more than the mere official sense the king of the situation" during the opening ceremony.[23] Of the same fair, the *Pennsylvania School Journal* noted that the button event "illustrated, as scarcely anything else could do this, the marvelous way in which the genius of man has tamed the forces of nature to his uses."[24] Examples of this sort pervaded popular discourses, where authors would describe the male button presser as a kingly master and the act as constitutive of man's control over his environment.

Quite unlike the descriptions common to US presidents, those written about presidents' wives received a specific kind of gendered attention in press accounts of these events. Here, women's perceived gentility, frailty, and inexperience were emphasized to stress the need for button pushing as a form of augmentation. The *North American* paper of Philadelphia (1886) related to readers how Mrs. Cleveland readied to push the button to start the Minneapolis Industrial Exhibition from the White House, noting that she was "comfortably attired in a white muslin dress, belted with a sash of delicate pink." Describing the event, the article continued, "Mrs. Cleveland stepped forward to give the signal which should move the machinery more than a thousand miles away. The spectators laughed heartily when the President gravely admonished her not to start it with a jerk."[25] Here, the newspaper emphasized Mrs. Cleveland's feminine and "delicate" appearance, which served as a reinforcement of the masculine nature of the button act by mentioning President Cleveland's admonishment. Although the president's wife received the honor of pushing the button, she quickly became the butt of a joke for her technological incompetence.

The *Atchison Daily Globe* (1888) described a similar scene when former President Polk's wife started the Cincinnati Exposition with a push. In addition to addressing the circumstances of the button-pushing event, the paper noted that Mrs. Polk "is described as a fine looking old lady, with white hair and erect, dignified carriage. Every year that she lives she becomes more notable from her connection through her husband."[26] As the story continued, it interjected details of Mrs. Polk's appearance into her involvement with the technological spectacle; presidents' acts often garnered descriptions merely in terms of their hands and their command of the button, whereas authors put women's bodies as a whole on display. Making hands and bodies visible for readers, news articles showed a woman at work, portraying her as relatable and domesticated while also in control.[27] In general, the world's fairs often served as breeding grounds for contestations about gender and technology, a setting where women inventors fought for rights in the face of increasing discrimination.[28] Women button pushers were often marked by their gender and could not escape it; their femininity featured prominently as a justification for push buttons to exist in the first place. Jokes and reports on button pushing, as in the case of Mrs. Cleveland, codified both appearance and performance to demonstrate the female button pusher's limitations. Although readers could not witness how Mrs. Cleveland pushed the button or what she looked like, the author imposed the button pusher's gendered body on the story.

In another instance, journalists speculated that US President Cleveland would allow his baby daughter, Marion, to touch the telegraph button that would inaugurate the Atlanta Exposition. Numerous reports commented on the potential affair by drawing attention to the baby's hands, with headlines (1895) such as "Her Dainty Touch," which made reference to "little Maid Marian" [sic] and the tiny fingers that could set a world's fair in motion.[29] Frequent references to weak, feminine hands made strong appeared in popular texts and reinforced this concept, as in the case of author "Miss Morning Glory," who wrote in her diary of a "Japanese girl" to this effect: "I look upon my finger wondering how such an Oriental little thing can make itself potent like the mighty thumb of Mr. Edison."[30] Using her finger as a marker of difference compared with Edison's "mighty" thumb, the author drew on a stereotypical rendering of Asian culture as submissive and delicate—yet made powerful by virtue of an American electrified button.

Over the years, these opening ceremonies via a button continued prominently, and the act of a finger push stimulating the wheels of a great machine from thousands of miles away also took on a metaphorical quality in newspaper accounts. An editorial eloquently captured the button's symbolism as a measure of the president's efficacy in office: "The other day President McKinley touched a button in Washington and in response the huge and complicated machinery in the great exposition at Nashville—700 miles away, awakened and started into active life. That is a symbol of the President's power if he but uses his high office in a way to convince his countrymen that he holds his place as a sacred trust and that his highest thoughts are for the welfare of the people."[31] The author later concluded, "We believe that in these days [President McKinley] is anxiously looking over the keyboard of the Nation and trying to select the button which will start the wheels of industry, and kindle anew the fires of hope. The button is there. God grant that he may find and touch it."[32] This striking metaphor of the president's position of power as calculated by his ability to select the right button and manipulate the nation from a distance suggests how pushing buttons remotely took hold as a compelling way of understanding governance as a kind of machine that one could effectively stir into action at a touch. If the president could touch a telegraph key to "awaken" machinery and bring a fair to life, then he should similarly have the ability to operate a metaphorical switchboard for the country's betterment.

As time passed, button-pushing spectacles for world's fairs became increasingly complex to showcase the spectacle of touch creating effects at a distance. No other tele-operated event would match the one designed by fair organizers to start the 1901 Pan-American Exposition in Buffalo, New York. The event hinged on an ambitious scheme to have US President William McKinley and other presidents from across the Western Hemisphere touch buttons at the same time. Complicating matters, President McKinley would travel cross-country on a train while pressing his button. This elaborate performance capitalized on a fantasy of remote effects:

At 2 o'clock, Buffalo time, by arrangement with the cable companies leading to South America, and with the telegraphic companies, and with the Atlantic cable companies, the Presidents and Rulers of all the countries of the Western Hemisphere will be requested to touch an electric button in their office, which will thus start a piece of machinery of the Exposition. ... President McKinley, from his special train, will start the great fountain pumps, and will transmit over the wires a message of greeting to the people assembled on the occasion of the opening.[33]

For those attending the exposition, these button touches—and the many mechanisms and people required to make such an arrangement possible—would have occurred far from view, making the seemingly spontaneous animation of machinery all the more spectacular. Newspaper reports on the coordinated spectacle called this a "novel" and "remarkable" plan, emphasizing how rulers and nations could come together through the use of technology to demonstrate their combined power. This striking example of collaborative work across great distances—where cable companies, telegraphic companies, and locomotive engineers merged around a common purpose—received voluminous praise across journalistic accounts for its efforts to coordinate action from a distance. Whether riding on a train, sitting in the White House, or standing on a fair's podium, a world leader could set his finger on the button and bring an event to life, these sources noted, pointing to a kind of omnipotence once unimaginable.

Just as media stories could act as sutures between a button-pressing event and a fair event, they could also work to expose the proceedings as artificial—and inauthentic—when a president didn't actually control the technology. In most cases, fair organizers would have wires strung from the White House to the fairgrounds so that telegraph and machinery were directly connected, but in the case of the 1904 St. Louis Exposition, President Roosevelt only sent a signal and an indirect one at that. The *Boston Daily* reported on this faked performance in an article titled, "Not by President," in which the author described a plot by the fair's president, David R. Francis, to simulate the button act: "President Roosevelt had telegraphed Pres [*sic*] Francis earlier that he had important engagements that would call him away from the White House chamber shortly after 1 and that he would be compelled to touch the button soon after 1," the author noted. "Pres Francis on receiving this looked worried. ... When President Roosevelt's message was received to start the machinery and the cascades, Francis ignored it till Taft had made his address. He then pretended to receive the signal and the machinery and waterfalls were started while the flags burst forth from the exhibit palaces. No one in the crowd was the wiser."[34] While fair organizers worked on the ground to keep the spectacle intact—taking advantage of the president's remoteness—reporters offered an exposé on the actual, fabricated goings on between fair organizers and the president. Although media accounts often reinforced dominant narratives of presidential control and national pride, they could also provide insight into the more mischievous and inauthentic sides of the function by acting in a watchdog capacity. Although the long-distance event could succeed on the ground, journalists intervened by weighing in on the act's authenticity for readers.

While these widely discussed and attended celebrations of push-button effects occurred on a national stage, hobbyists and young experiments played with buttons that promised a thrill while discreetly hiding away the source of control. Myriad catalogs and magazines advertised novelties for purchase that would offer their users inexpensive opportunities to explore electrical effects. Chief among these included toys for practical jokes and pranks that took advantage of unwitting spectators and participants, providing a shocking—both surprising and painful—form of interaction. As with other kinds of lessons about push buttons, advertisers often targeted these novelties at boys—in contrast to how girls became showpieces for effortlessness—and framed them as opportunities for thrilling amusement. Their affordable price and accessibility to a wide range of consumers made these practical jokes one of the earliest sources of experimentation with push buttons for laypersons.

Indeed, the *Sentinel* newspaper of Milwaukee (1887) defined a "practical joker" as "the fellow who has a rubber nail in his office to hang coats on, a dummy electric button with a needle to call for a messenger; who coats the gum of his envelopes with red pepper, and who tells you he can buy thirteen two-cent stamps for a cent and a quarter."[35] In this description, the paper attributed both harmless pranks (rubber coat hanger) and painful encounters (needle masquerading as electric button) to the practical joker; across the board, the performer occupied a position of power by controlling those not privy to the joke and demonstrating his mastery over electricity.

One of the most common electric button jokes played on two technologies with which many people would already have familiarity—the chestnut bell and the electric bell. The former, a bell worn on a waistcoat and rung in mocking whenever someone told a story or bad joke one too many times, provided an increasingly cruel and sophisticated form of entertainment when combined with electricity: "When you have one on, and someone tells you that your favorite story is a chestnut, you suavely tell him to ring it up for you," the San Francisco *Evening Bulletin* instructed readers about the electric chestnut button. The article recommended, "As soon as he presses the chestnut button a needle point runs into his finger and announces that the laugh is on him."[36] A New Jersey catalog for electrical novelties and other supplies similarly promised that the electric button, sold for 10 cents, was "Just the thing to use on old chestnut stale-joke peddlers. One shock from the Electric Button and he will never try to 'rim off' any chestnuts on you again, but he will immediately send for a button himself, *and square his account with some other fellow*" (italics original).[37] Producers sold the electric chestnut button as a playful "improvement" on a popular gag while reinforcing the notion that the button's owner could employ the novelty to serve a disciplinary function that warded against unfunny jokes. One shock, advertisers encouraged, would dissuade the use of these jokes and enable the former victim to take his own revenge on the next unsuspecting participant.[38] This pay-it-forward push certainly promoted tomfoolery and amusement with electricity, but it also lent an incredible aura of power to push buttons by making them impenetrably opaque and surprising. A seemingly safe and simple push of a button—whose design meant that pushes did not get any feedback or instruction before touching—could at once become the trigger for humiliation and physical pain on the part of the unwitting pusher. Like the familiar chestnut bell, novelty versions of electric doorbells were made for purchasers to wear on their clothing. These toys made fools of the curious, trusting spectator who possessed an interest in new gadgets while playing on the psychology that anyone offered a button would press it. An ad (1887) offered "The Electric Button" for 15 cents, a device that "looks very tempting, and attracts the curiosity to PUSH it, which never fails to produce a shock that will make them dance."[39]

Another form of push-button electric magic designed for purchase involved a button-activated electric light for wearing on scarves, neckties, and lapels. The Stout-Meadowcraft Company of New York (1885) began advertising the "electric light scarf pin" for $5.00. The pin operated by affixing a lamp to one's clothing that, when connected to a silk-covered cord, battery, and a push button hidden in a pocket, could bring a spark of electricity to common attire (see figure 4.3). The company promised potential consumers that "the most sensational and amusing effects may be produced by this scarf pin, bewildering and surprising those who are not in on the secret, as there is nothing apparent which would indicate the manner of producing the light."[40] More than the spectacle of wearing an electric light on one's clothing, the advertisement emphasized the device's secret method of control—of concealing the button beneath one's clothing—as the intriguing and surprising element of the electrical adornment. By creating distance between action and effect, the novelty promoted push buttons as conveying mediums for producing electricity at will and with simple convenience. Users did not even need to see the switch to operate it.

Over time, the price of this electrical novelty had greatly decreased, and Ohio Electric Works of Cleveland (1897) lauded

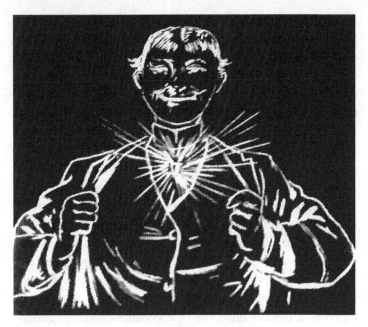

Figure 4.3
Individuals wore electric scarf pins to adorn their bodies with electricity, sometimes to demonstrate its surprising and magical effects, and sometimes for practical purposes of employing portable light.
Source: Ohio Electric Works, "An Electric Light for the Necktie," February 8, 1897. Image courtesy of the Warshaw Collection of Business Americana—Electricity, Archives Center, National Museum of American History, Smithsonian Institution.

the invention for its evolution in an advertisement: "An idea of the wonderful progress in electrical science is well demonstrated in the perfection of this necktie light," the company boasted. "But a few years ago it was considered wonderful when an electric light of any efficiency was produced, even by costly and cumbersome machinery; now we use it to adorn the person and generate electricity in the vest pocket by an instrument cheaper and more reliable than a watch."[41] The ad also noted that consumers could purchase these lights in any color, clear crystal, or opal, which "appears like a 'ball of fire.'"

The new-and-improved scarf pin light at the end of the nineteenth century, worn primarily by boys and men (according to the ad), exemplified strides made in the areas of electrical control. Comparing the light to a commonly accessible watch, Ohio Electric Works sold a vision of adornment with electricity easily put into the hands of its user. The "novel" button in a breast pocket, tucked safely away from the site of the spectacle, could give rise to a brilliant "ball of fire" sure to impress neophytes unfamiliar with the particularities of electrical mechanisms. The same company also offered a pricey $3 version of its light for a necktie or coat, proclaiming its popular use among "actors, watchmen, mail carriers, messengers, milkmen, and others." More than a frivolous amusement, the light served a practical function of lighting one's way in occupations typically characterized by nighttime work or in other places with darkness.[42]

Other instructions for a host of remote electrical curiosities made possible by button pushing circulated in tomes such as R. A. R. Bennett's *How to Make Electrical Machines* (1902), which offered elaborate designs for boys to create a doll that could play an electric trumpet without human hands ever interfering. Wrote Bennett of this novelty: "Hide the battery in a corner in a black box, the wires coming through the side next the wall, and the press in a dark corner, or on the floor under a table so that you can put your foot on it while your hands are free, writing, etc. You can of course now tell the doll to blow, at the same moment putting your foot on the press, when the trumpet blows accordingly."[43] The author noted that, much to the inventor's delight, "Of course this is mysterious to the last degree to the uninitiated friend to whom you are displaying the doll, as you may be any distance off from the doll with your hands free, speaking to him across the room."[44] This description emphasized, like the electric button shock novelty, how to fool the "uninitiated" spectator by separating electrical act from human body. By stepping on a button situated across the room, a seemingly inanimate doll could come to life and blow a horn without any evidence of the button presser's involvement. These push-button acts offered opportunities to explore and engage with electricity in ways that made the intangible seemingly tangible and the effect far more dramatic than the mere touch that caused it.

Take an ad for an electric money bank, for example, which offered elaborate instructions on how to fool a well-meaning patron who wished to contribute money to a child's bank: "When company comes to the house bring out the Bank and they will read the words on the top and drop a penny in it every time, and when they push the Button, you will see some pretty lively Dancing right away," the piece instructed. Promoting the fun of surprise and bodily stimulation, the ad concluded, "The dancing will be done by the one who pushes the button, as they will receive a shock that they will not forget very soon."[45]

The advertisement emphasized how the toy bank, once a neutral object that its owner could show off to visitors, transformed into the conveyor of an electrical (and supposedly unforgettable) performance. A button stood at the center of this experience; one could not turn back from "dancing" once initiating a push (see figure 4.4). Experiments with giving and getting electric shocks were numerous during this time period, and well-known gags such as the "electric girl" at the county fair whose hand would dispense a shock from her body when touched while "defeat[ing] enquiry" played on the dramatization of secrecy,

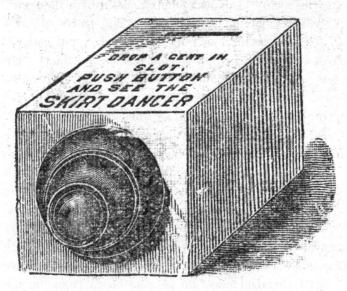

Figure 4.4
The electric bank, a shocking novelty, took advantage of the unwitting participant who may not have much familiarity with electricity or push buttons.
Source: Ardee Manufacturing Co., *Illustrated Catalogue: Manufacturers, Importers and Jobbers of Toys, Novelties and Mail Order Merchandise* (ca. 1903 or later). Image courtesy of Trade Catalog Collection, Baker Library, Harvard Business School.

magic, and sensory experience.[46] Author George M. Hopkins went so far as to proclaim in his book, *Home Mechanics for Amateurs* (1903), that "the giving of electric shocks to one's friends is always a pleasant pastime," encouraging young experimenters to literally take electricity into their own hands.[47] Children's pranks, in the context of domestic environments and on streets, offered initiation for youth into a culture of scientific wonderment and exploration, where the child could have control over the unschooled and unfamiliar adult.

Not only did push-button mischief play with the dangerous, surprising, and concealed nature of electricity, but these pranks also fit contextually within the late nineteenth-century, which has been described as an "age of puns, parodies, quips, hoaxes ... practical jokes."[48] Men—called "jolly fellows"—were typically responsible for this kind of carousing and humor, celebrating masculinity and even violence as part of their repertoire.

When it came to users' feelings about the potency of fingers pushing buttons to generate effects across distance, opinions existed at both ends of the spectrum, either attributing a tremendous sense of power to hands or describing them as largely impotent. Poet and philosopher George Woodward Warder (1901) believed that "[God] touched the electric button that gave impulse to all atoms, created all suns, evolved all worlds, and sent them singing in harmonious motion through all space, for all eternities."[49] In religious discourses, the "finger of god" had long featured prominently, associating hands with creation as well as destruction. For, according to one sermon, "We are under [God's] eye and his finger; and one look of that eye could blast us, one touch of that finger could crush us."[50] The cause-and-effect and all-or-nothing functions of push buttons further facilitated such a view of whole worlds brought into

being or exterminated with a push. In this regard, button pushers' fingers functioned in a forceful and godlike manner, and
with agency.

Thinking through this relationship between button pushers
and the effects of their pushes, one writer similarly drew on religious rhetoric: "Deeds of grandeur or deeds of terror are accomplished with less immediate effort, and at a distance from their
effect. The touch of a button executes a murderer or starts all
the enginery of the Columbian Exposition. Is not this somewhat
the way that God works?"[51] This capacity to spur people and
machines into action at a far remove seemed to promise the best
of the electrical age: liveness, connectivity, visibility, safety, spectacle, or even the power of a god. Writers who perceived button
pushers as godlike imagined that touch could have a great ripple
effect across the country or even the world, and this form of
remote control inspired an unrelenting pursuit of digital command in the name of humans' taming of electrical forces.

Yet another writer described the "spiritual" experience of distant effects through electricity. Author Gerald Stanley Lee (1901)
argued to this effect in an ode to machines that, "Every time
[man] touches a material thing, in proportion as he touches it
mightily he brings out inner light in it. He spiritualizes it." He
proposed, for example, that a man of the Industrial Age, rather
than using a door knocker of the past, "likes it better, by touching a button, to have a door-bell rung for him by a couple of metals down in his cellar chewing each other. He likes to reach down
twelve flights of stairs with a thrill on a wire and open his front
door."[52] He described how someone could figuratively "reach"
through wires, once again connecting the value of reachability
to this form of digital command made possible by push-button
electrical control.[53]

One striking case of thinking about this potency of reach and distant effects involved applying the bell-ringing metaphor to medicine; physicians often made the argument that parts of the body worked like buttons, where a push in one place could stir up effects in another. For author Emma Curtis Hopkins (1894), thoughts in the brain represented touches of a button that would "ring" throughout the mind and body; bad or "discordant" thoughts would send that body into disorder.[54] For another author, the whole sympathetic nervous system was made up of "thousands of little electric buttons" and "innumerable switch-boards" that communicated with one another (see figure 4.5).[55] Yet another suggested that digestion might fail if the push button were broken, whereas a functional button would send messages from the reflex center to one's nerves.[56] Physician George William Winterburn (1900) concluded that disease compared to the mechanism of an electric bell, in which "pressure on the push-button closes the circuit" and caused bacteria to circulate.[57] Similarly, Andrew Taylor Still (1910) theorized that life came from touching a button that would cause one's heart to beat and to generate electricity throughout the body.[58] Over and over again, medical professionals used push-button analogies not only because they made for attractive images and mental models in sync with electrification efforts of the time period, but also because they referenced an everyday, relatable technical object that epitomized communication and control through straight-forward cause and effect. Buttons represented how a single touch in one location could rapidly travel across distance with tangible and often profound results in a kind of domino effect.

In perhaps the most significant instance of button as metaphor for distant effects, in a widely cited article about female circumcision, Dr. Robert T. Morris (1892) proclaimed famously

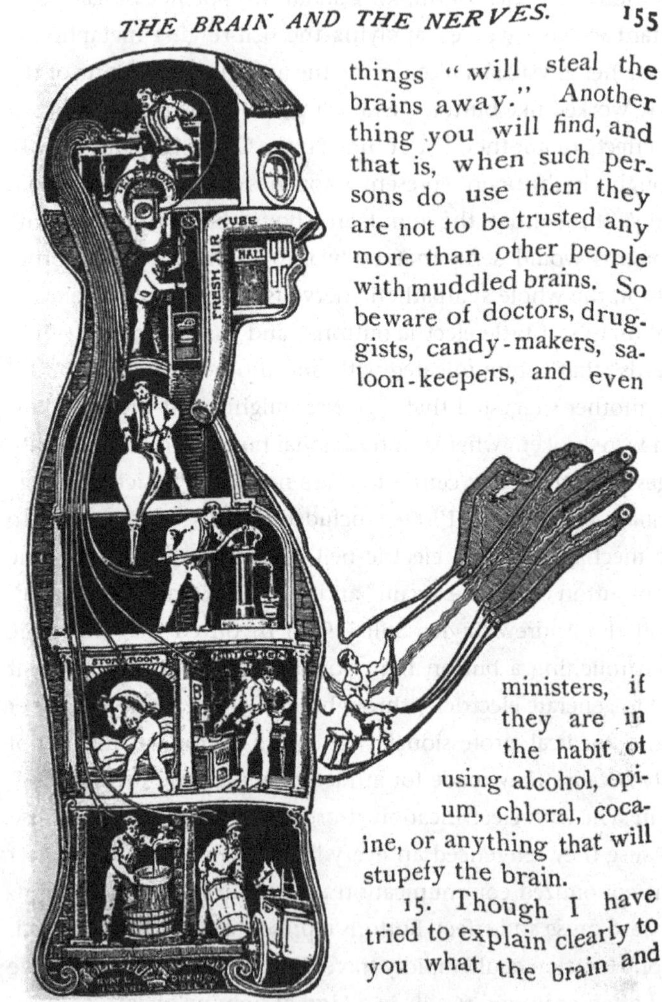

THE BRAIN AND THE NERVES.   155

things "will steal the brains away." Another thing you will find, and that is, when such persons do use them they are not to be trusted any more than other people with muddled brains. So beware of doctors, druggists, candy-makers, saloon-keepers, and even ministers, if they are in the habit of using alcohol, opium, chloral, cocaine, or anything that will stupefy the brain.

15. Though I have tried to explain clearly to you what the brain and

Figure 4.5
Depiction of fingertips as buttons activated by the body.
Source: Jerome Walker, *Health Lessons: A Primary Book* (New York: American Book Company, 1887). Image courtesy of University of Illinois at Urbana-Champaign Library via Google Books.

to the medical world that, "The clitoris is a little electric button which, pressed by adhesions, rings up the whole nervous
system."[59] The doctor proposed that this malfunctioning "button," which could control a vast network of physical symptoms
and effects (ranging from epilepsy to nervous disorders) from
a distance, must be treated to restore a woman's out-of-control
system back to health. This call for female circumcision attracted
much discussion within the gynecological and obstetrics community, with many voicing support for Morris's conclusion.[60] As
a vivid metaphor for the way that women's bodies worked (or
how some believed they worked), the electric button construct
served to demonstrate how the medical establishment could
control feminine anatomy and women more generally by viewing them as machines set into motion by a single touch; this
approach fit within broader mainstream societal efforts to medi-
calize women's sexual arousal and label it as a "crisis of illness."[61]
This crisis manifested from, in the words of physician Benjamin
E. Dawson, who supported Morris's conclusion, the "electric
push button, which from irritation it may ring up disastrous
reflexes in remote parts of the body or transform a healthy sexuality into a jangling sensuality."[62] An "out-of-order" push button
created a condition that required physicians' interventions.

Without question, individuals employed metaphors of the
body as a machine well before Morris and continued to do so
afterward.[63] The fact that the doctor chose an electric button
as his metaphor and that this metaphor perpetuated, however,
suggests specifically how physicians imagined women's bodies
as controllable and buttons as dangerous sources of activation.
It also demonstrated a close interrelation among femininity,
mere touch, and push buttons. The image of a button conjured
a cultural association of swift action and reaction, of potent yet

simple electrical control, and the speed and intensity of this control with a single touch. Just as important, it identified a power differential between button pusher and the "pushed" that allowed for easily justified domination. Doctors strove to fix the "button" so they could restore balance to the body to maintain social control. Women thus existed in a kind of limbo position as desiring subjects—at once made into spectacles as ideal consumers of pleasure—and yet demonized for their sexual desire. The push-button metaphor of women's sexual organs paved the way for future thinking about buttons in relation to sexuality, referring commonly to "turning someone on" by pushing her buttons.

The sexual effect metaphor of the clitoris described buttons as the harbingers of disastrous or undesirable effects, which required controlling or eliminating the button. When it came to electricity, seemingly unstoppable and instantaneous results swiftly rendered also produced anxiety. Although push-button bells typically worked by momentary action—they only rang when a finger stayed on the button and stopped when the finger lifted—what of other buttons that might start but never stop? How to make sense, for example, of future prophecies of doomsday buttons? Could a button pusher, unable or unwilling to turn back, end the whole world at a touch? When button pushes were perceived as irrevocable and dangerous, carrying out action at a far remove from the finger that animated them, their utility and social significance took on a different tenor.

Indeed, as early as 1892—more than 50 years before the political anxieties generated by the push-button warfare of the Cold War era—pushing buttons came to symbolize a fear of long-distance, instantaneous warfare. Although push-button warfare existed only in popular imagination as portents of a future where button pushing could end the lives of everyone in a country, or even the world, in these prophecies, an all-powerful button pusher could enact swift, irreparable effects with a single press by setting unstoppable forces in motion. In an article titled "The End of War," author J. F. Sullivan envisioned a world where "war seemed to grow ever more terrible; until it came to such a pass that a single human being could destroy a whole nation by simply pressing a small button with his finger" (see figure 4.6).[64] As Sullivan imagined, it was not a crazed dictator or power-hungry politician who blew up the world, but rather a bumbling gentleman who unwittingly and effortlessly pushed a button that he happened to encounter without realizing what effects the button would trigger. This extreme case—holding an entire nation's fate in one's hand (or rather at the touch of one's finger)—brought into question views about power, control, communication, and effortless machine interventions. At the same time, it suggested that buttons were too simple and swift in action without checks and balances in place to prevent the ignorant button pusher from bringing about ruin. The reachability enabled by digital command also produced anxiety—if anyone could gain access to a push button, how could society maintain control over the unskilled, incompetent, or evil-doing controller?

To this end, because of buttons' simplicity and the way they initiated cause and effect, fiction writers also viewed them as the mechanism that could trigger an apocalyptic ending for humanity if buttons fell into the wrong hands. One author imagined a scenario in which Thomas Edison stood at the apex of a conflict between Great Britain and the United States, possessing the electric power to eradicate whole countries from the map: "In order to avert future trouble," the fictional Edison proclaimed, "I think

"A SINGLE HUMAN BEING COULD DESTROY A WHOLE NATION.

Figure 4.6
Prognosticators worried about future control over warfare. The editorial depicts one extreme example, in which an unwitting figure stumbles across a button and accidentally destroys a whole nation.
Source: J. F. Sullivan, "The End of War," *Strand Magazine* 3 (June 1892): 646. Image courtesy of Cornell University Library via Google Books.

it would be best to destroy England altogether."[65] After instructing his assistant to touch button number four, which obliterated the country, Edison concluded, "If we should ever be at war with any other nation, you have only to notify me. I have an electric button connecting with every foreign country which will destroy it when pressed. In ten minutes I could destroy every country in the world, the United States included."[66] This bleak portrait of Edison as an all-too-powerful scientist at a switchboard of buttons controlling the world spoke to the antibutton contingent's greatest fears. In such a scenario, with decision making localized in one person's hands and the ability to change the course of human history with so little effort, this writer and others imagined the push button at the center of the race's ruin.[67] Buttons conjured fears of all-or-nothing actions that could spiral out of control. Although this kind of catastrophic war did not exist at that moment, depictions of it presciently foretold the paranoia that would accompany a world where geopolitics rested on a push or press at the hands of a digital commander. Moral watchdogs frequently spoke out about this kind of warfare, whether malicious or unintentional.

Pushing a button to blow up a mine, start world's fair machinery, or initiate warfare each constituted an example of buttons as tools of activation. Thinking about pushing a button to set these forces into motion meant to confront an essential paradox, according to scientist Julius Robert Mayer:

Human nature is such that people like to achieve the greatest effects with the smallest possible means. The pleasure we take in firing a weapon is an eloquent example of this. ... But even if activating things is an inexhaustible source of permissible joy and harmless pleasure, we must also note that this phenomenon can also lead to the most heinous crimes.[68]

Where pushing a button certainly connoted effortless control, Mayer identified a theme common to the volatile nature of button pushing as a reversal of forces in the late nineteenth and early twentieth centuries; the forceless force of a finger touch combined with a catastrophic or far-reaching effect could carry with it grave and sometimes irreparable—whether imagined or not—consequences. In part, this viewpoint reflected a discomfort with carrying out an action that triggered results which couldn't be viewed by the button pusher. Scholar and author George Herbert Palmer (1903) wrote to this effect, "When I touch a button, great ships are loaded on the opposite side of the earth and cross the intervening oceans to work the bidding of a person they have never seen."[69] Where users perceived buttons as erasing distance in some circumstances, they viewed them as perpetuators of distance in others.

Social debates routinely occurred about the morality of technological interventions from a distance: what ethical guidelines should one take into account when pressing a button thousands of miles away that could take the life of someone else? Should matters of life and death take place only in face-to-face contexts? Dr. A. R. Wallace (1895), writing on examples of immorality and morality both hypothetical and real, concluded bleakly about the human condition that nothing "would restrain a poor, selfish, and naturally unsympathetic man from pressing the electric button which would at once destroy an unknown millionaire and make the agent of his destruction the honoured inheritor of his wealth."[70] Wallace believed that in times of desperation, anyone would hide behind the push button's shelter of anonymity and take the life of another for his own gain. Here, push buttons acted as simplistic vehicles for getting away with murder. Onlookers feared that localizing control to an instant, decisive, and remote touch—made possible by the invisibility of the effects—would provoke humanity's darkest impulses.[71]

These fears, although grandiose in their description, were founded in material changes occurring in military technology. Of note, those observing increasingly mechanical warfare described a moral as well as a manual shift in the physical act of carrying out war. According to Charles Morris (1898), "A modern battle-ship has grown to be an automatic machine, an instrument of warfare in which nothing is done by hand."[72] He remarked, "Now the hand has little to do, except to move levers, press electric buttons, open and close throttles, and the like."[73] Echoing sentiments common across industries, Morris contrasted manual labor from digital command—the minimalist intervention of hands pressing buttons seemed not to qualify as hand work, despite the fact that all of these controls would have necessitated routine gestures and adjustments to keep the battleship functioning appropriately.

This problem of the hand having "little to do" produced concern—and even outright indignation—in the late nineteenth century, when it came to matters of life and death. In particular, debates over the electric death penalty demonstrated a fundamental uneasiness with taking a life in a manner quite "'insulated' and at a remove from the body in question. In an essay on "Electric Killing" (1888), Thos. D. Lockwood noted that, although electricity could allow people to communicate in myriad ways, "we have not yet pressed a key or a push-button for the deliberate purpose of killing anybody."[74] Lockwood wrote in response to a suggestion made in New York, outlined in a report from the Gerry Commission to begin using electricity for the death penalty, and he called this a "cold blooded proposition for the degradation of a noble science; and moreover one which is entirely uncalled for."[75] Electricians convening at the National Electric Light Association Convention in the following year expressed similar concerns in a panel assessing the constitutionality of electrical capital punishment. Vocal dissenter Professor Anthony argued that no sheriff would want to "place the electrodes and touch the button which was to produce death."[76] He wagered that even after 100 years, electricians or other experts would still have to carry out the act because no nonexpert would take on such a weighty responsibility. Despite these outspoken rejoinders, only five months later, New York passed the Electric Execution Act, which conjectured that applying a scientific and technological method to executions would reduce public outrage toward capital punishment.[77] Given the "gentle pressure on the button" required to carry out executions, some also viewed the shift as a progression to "moral and intellectual rather than physical" ground, quite unlike hanging, stoning, beheading, or other more viscerally violent methods.[78] Public interest in the death penalty crystallized around this modern, technologically superior and masterful form of control that could take a life.[79] Push buttons combined with the electric chair to create a standardized, state-sanctioned form of justice that was "instantaneous" and "well calculated to inspire terror."[80]

Removing labor from the operation via electric button provided a useful justification for managing deviant behavior while making the activity less overtly gruesome and brutal. Yet some worried that people did not know enough about electricity to use it effectively for execution, nor should the criminal "be put out of the way in the easiest possible manner for him."[81] Death penalty by button could be perceived as too simplistic and therefore unfit for its weighty task, given that it operated with the same touch as an electric bell push; from this perspective, a "reversal of forces" violated a tenet that human life should not be taken without effort.[82] The binary nature of electric death—life and death tethered to a switch—also produced mixed reactions. Describing how prisons would carry out electrical executions, Thomas Edison—who originally opposed the death penalty before taking a financial interest in the process—relied on the push button as the key mechanism in carrying out the deed.[83] Edison famously remarked of this process, "When the time comes, touch a button, close the circuit, and," he said with a snap of his fingers, "it is over."[84]

In reality, however, early forays into electrical executions often did not conform to the on/off binary that Edison and others promised. The first execution by electric chair—of prisoner William Kemmler—produced disastrous results when 17 seconds of applied electric current did not take the man's life as expected. According to a *New York Times* (1891) article reporting on a subsequent death penalty case that would use the same method, the electricians responsible for the failure "have not forgotten that life apparently came back to Kemmler after it was thought that he was dead and the current was turned off."[85] Far from the "snap" that could toggle between life and death, the realities of electrical capital punishment suggested that execution by button required an array of technical forces to come together successfully. In the ensuing years, push-button executions evolved into routine practice, viewed by many as a more humane way of killing than those of the past. By transforming violent physical actions into mere touch, push buttons stripped physical force from the death penalty act while leaving the forceful impact of death in its wake.

Although buttons sometimes functioned differently from how they were imagined in electrical execution, many continued to invest in the notion that buttons could provide instantaneous and direct retribution for those wronged, balancing out the scales of good and evil. The *Washington Post* (1892) indeed reported that a father, after the death of his daughter, hoped that "I may be allowed to touch the button of the electric machine that kills the man that murdered my daughter."[86] Putting control in the hands of the victim, the push button could serve as a tool of empowerment, but it also raised societal concerns about what it meant to take a life with a single push. Those who advocated against button pushing believed human life was too sacred to take so easily. Pushing buttons prompted deep and sometimes unanswerable questions about society's impulse to carry out life-and-death decisions from a distance, to put control within one person's hands and under one person's finger.

Given the wide array of interpretations about generating push-button effects at a distance, it became difficult to pin down whether buttons were purveyors of pleasure, panic, or some strange mixture of both emotions. Just as a society in the midst of industrialization and electrification had to negotiate what it meant to communicate at longer distances through technologies such as telegraphs and telephones, so too did the question of action at a distance prompt negotiation over what forces button pushers might set into motion. Where setting people into motion could provoke frustration, when it came to electricity, this kind of activation often led to anxiety.

5 We Do the Rest

This is emphatically a fast age, mechanically as well as morally speaking, in which we are living. "You press the button and we'll do the rest" is a current phrase that characterizes about everything done nowadays, from the efforts of the camera fiend to the work of the car shop. Perhaps our push is out of proportion to our patience. At any rate, we can no longer endure the tedious, tortuous methods of our fathers, if any other way can be devised to get there quicker and easier. ...[1]

Some one aptly remarked the other day that this is a push-button age and that push-button engineering, as it might be called, has brought about an infinity of comfort and convenience for everybody. "You press the button; we do the rest," has, as we all know, been a familiar phrase for some time, and what it means is now so generally taken as a matter of course that we have ceased to wonder at the marvels which the simple push-button commands.[2]

The most extreme examples of button pushing, from inaugurating a world's fair to taking a person's life, served as demonstrations of humans' ability to "tame" electricity and bend it to their will, when these events transpired as planned. This construction differed in its emphasis on the potency of a finger, quite in contrast to the oft-noted inefficacy of a finger that rang for a person. Although these publicized and large-scale events may not have affected most Americans on a daily basis, electricians,

manufacturers, and advertisers worked to extend the romanticized concepts encompassed by digital command of electricity
to everyday life for consumers—beyond those who could afford
servants. To do so meant to conceive of ways that electrical
machines could deliver satisfying, instantaneous gratification
without human effort or fear of harm, and discussion shifted
from "taming" to the notion of perfect user control and perfectly controllable machines. Button pushing was thus situated
contextually within American discussions about automaticity,
and the term "automatic" often covered a wide variety of con-
voluted and contradictory meanings.[3] In reference to buttons, it
referred to a pervasive cultural craving for efficient relationships
between humans and machines. Buttons were lauded as the
"acme of convenience" by mitigating labor and complication
from electrical experiences, so that machines could take over the
lion's share of effort (see figure 5.1).[4]

To animate the working parts of a machine into motion or
start the flow of electricity at a touch involved thinking about
that machine or force as "automatic" and capable of completing tasks that required only a finger command. The electrical
industry depicted this kind of machine as a boon, but automaticity tremendously threatened and upended deeply ingrained
definitions of craft and skill by paving the way for a new breed
of hand, user, and consumer.

Otis Elevator Company was one of the first to adopt the
"automatic" moniker after years of manufacturing elevators
that operated by hand cranks or "hoists." The company initially installed slow-speed, push-button–controlled transportation devices in residences and apartment houses for consumer
use.[5] Average elevators of this type cost approximately $3,000
apiece; because of their expense, only luxury homes or urban,

THE "AUTOMATIC."

Figure 5.1
Advertisement for an "automatic" push-button switch made to appeal to an elite class of early electricity users.
Source: J. H. Bunnell & Co., "Illustrated Catalogue and Price List of Telegraphic, Electrical & Telephone Supplies" 9 (1888): 137. Image courtesy of the Warshaw Collection of Business Americana—Electricity, Archives Center, National Museum of American History, Smithsonian Institution.

multifamily dwellings would most commonly install them.[6] Frank J. Sprague, one of the inventors of the electric elevator that debuted in 1888, described how the earliest elevator buttons worked: "If the machine is at rest, the pressing of a button calls the car, wherever it may be, to the particular floor at which it is wanted, where it automatically stops."[7] Once inside the elevator, different kinds of button configurations existed. A two-button system would have buttons for "Up" and "Down" and would require the passenger to continue pressing the button while moving, doing her best to land evenly with the floor she desired. A three-button system would function similarly, with the exception that the user need not hold down the directional button to move and instead would press the third switch to halt the elevator's movement while trying to guess the location of the floor desired. The last system, most congruous with contemporary elevators' functionality, provided a different button for each floor, and the car would stop automatically at the place requested without further action on the part of the passenger.[8] In its advertising for this kind of elevator, Herzog Telesme Company emphasized the simple notion of "up" and "down" at work in its product, with copy that instructed users: "Take me UP/Push the Proper Button/This Shows in All Cars/The First Available Car will Stop to Take You/Take me DOWN" (see figure 5.2).[9] Condensing the complicated mechanisms of elevators into binary choices, up or down, push-button elevators were designed to remove complexity, labor, and the extra set of hands typically associated with the technology. They opened the market of vertical transportation to a new class of passengers who desired luxury, comfort, and convenience; before long, homebuilders had deemed the automatic elevator a "necessary adjunct of the modern city house."[10] Users could operate any elevator by "merely pressing the button," where "literally the only mechanical skill required for operating is the ability to push a button."[11] These common assumptions about elevators echoed popular rhetoric of simplicity, effortlessness, and no requisite skills on the part of users. Proponents of push-button mechanisms argued that a touch of a button would allow physical beings to transcend the limitations of their bodies. They imagined that any human body—whether ill, weak, or simply inexperienced—could act as a powerful agent, a digital commander, and this image of push buttons fit into a broader cultural ambition at the turn of the twentieth century that machines could make bodies better, stronger, and more socially productive.[12]

Indeed, advertisements for Otis elevators called on this familiar trope by suggesting that "a child can operate the Otis Elevator."[13] Accompanying images to this slogan typically featured a young girl in a well-appointed residential elevator, standing just a bit on her toes to push a button with a slight smile on her face (see figure 5.3). Although elevator operation did not involve carrying out an act of destruction, like blowing up a mine in the case of Mary Newton, the idealized Otis girl drew attention to the fact that even the smallest, seemingly undeveloped members of society could have access to the greatest machine when empowered by a push button. Indeed, in the words of Otis, "[The elevator] can be operated by the oldest or youngest" member of the household.[14] If age did not play a significant role in elevator operation according to these push-button evangelists, then neither did skill because, according to one history of science book, "literally the only mechanical skill required for operating is the ability to push a button."[15]

When selling button-powered elevators in forums typically targeted at architects, electricians, and builders, the Otis

Figure 5.2
Labeling for elevators that emphasized the binary control mechanism
for laypersons.
Source: Herzog Telesme Co., n.d. Image courtesy of the Warshaw Collection of Business Americana—Electricity, Archives Center, National Museum of American History, Smithsonian Institution.

OTIS
Electric Elevators

ELECTRIC Elevators are particularly well adapted for private residence service. With our improved Push Button System of Control, a regular attendant is unnecessary. The operating device so simple that a child can operate the car with entire safety.

The machine can be operated both from the car and hallways.

By pressing a button placed in a hallway the car can be brought to that floor, stopping automatically when opposite the landing.

Figure 5.3

An Otis Elevator Co. advertisement drew upon images of young children operating elevators to emphasize their simplicity of operation and lack of effort required.

Source: Otis Elevator Co., "Otis Elevators," *Good Housekeeping* (1908): n.p. Image courtesy of University of Iowa Library via Google Books.

Corporation focused on three tactics: (a) emphasizing the elevator's ease of operation, (b) demonstrating the device's safety mechanisms in place, and (c) outlining how push buttons would provide "perfect" control to their users.[16] Inventors particularly touted binary control—and management over when buttons were touchable and untouchable—as safety features: "When the door is open the electrical circuit is broken, making all buttons inoperative, and holding the car immovable at that floor until the door is securely closed and the button pressed."[17] Limiting the user's ability to push buttons created parameters around pushing to keep fingers from performing out of bounds.

Importantly, the shift to automatic push-button control of elevators both explicitly and implicitly involved a movement toward elevators that could operate without an attendant.[18] Although it would take well into the twentieth century before users primarily operated elevators, elevator manufacturers at an early stage began invoking the image of elevators as willing servants who responded to anyone's touch—not just that of the elevator operator: "It is only necessary to press an up- or down-hall call button once and the car will always respond to the call."[19] Buttons figured prominently in arguments for doing away with designated elevator operators and allowing electricity to perform an act of willing servitude. Companies such as Otis promised that the technological elevator attendant would not respond begrudgingly to a summoning request as did many human beings in similar positions; instead they would "always respond to the call."

The push-button elevator concept related to other human–machine experiments that were designed to put consumers in control of their own consumption by eliminating intermediaries like store clerks, elevator operators, bellboys, and others. Thomas Edison, for one, forecasted a future in which "The Automatic Store" would reign supreme.[20] Edison—who had advocated for burglar alarms, electrical execution, and numerous other button-powered devices—surmised that "clerkless shops," run entirely on credit and without a human element, would make vending machines the centerpiece of a new form of commerce. The renowned inventor believed that consumers, through pushing buttons, might access anything they desired at a moment's notice. This concept first came to fruition when the United States granted its first vending machine patent in 1884.[21] Vending apparatuses, many of which became highly profitable, fit into broader patterns of recreation and leisure spending common in the nineteenth and twentieth centuries that privileged "mass" consumption of inexpensive entertainment and products.[22]

In addition to offering opportunities to witness spectacular effects, world's fairs and expositions served as forums where attendees could interact with early versions of electrical vending novelties that may have been unavailable to them in everyday life. In July 1892, nearly nine months before the start of the World's Columbian Exposition in Chicago, the *Daily Inter Ocean* newspaper warned potential fair participants about the many attractive push-button activities that could put a strain on their wallets. Titled "Another Button Scheme," the article claimed, "Visitors at the fair next year must be careful about pressing the button, otherwise they will find a constant demand upon them for money."[23] Despite the author's wariness, the article noted, "The button is to be one of the great features of the fair." Describing the push-button activities available to patrons, the article continued:

One button when pressed will return a glass of ice water, another will throw out a sample of chewing gum. ... In the German village it will only be necessary to press a button and a fresh, rosy-cheeked and buxom fraulein will appear with a tankard of foaming lager. In the "Streets of Paris" one touch on the button will bring most anything the visitor may want, from a bottle of wine to a perfumed bottle.[24]

The article's imagining of the push-button world available to fair participants dramatized an experience based on instant gratification where buttons delivered anything one could desire, from a woman to a bottle of wine. Still, in the paper's encouragement of these novelties and its proclamation of the button's role in the fair, a note of ambivalence existed about the cost of these entertainments. The author viewed push buttons as a marvel—not practical necessities—and thus patrons must guard against getting taken advantage of and investing too highly in this gimmick. Here, and in many other moments, buttons got caught up in contradictory rhetoric among utopian control technologies, whimsical amusements, and ordinary technologies.[25] Events like world's fairs helped to associate push buttons with consumption and the fulfillment of one's desires; they encouraged novices to explore and embrace buttons in a celebration of technological progress and consumerism.

World's fairs offered convenient test cases for push-button consumption in a second sense, too, in that they could provide inventors with information about what could go wrong with buttons. In particular, some children found amusement in toying with early vending machines, much as they did with electric bells. Author Arthur F. Wines (1899) noted that children at these fairs had infiltrated candy machines, leading to critical changes made in the ensuing years: "Boys discovered that by careful manipulation of the push-button it was easy to determine the critical point at which the confectionery would be ejected, and that by setting the thumb hard on the button so as to form a fixed stop, they could empty the machine as rapidly as the hand could operate it."[26] Methods of "fraud" and various ways to "beat" slot machines persisted before countermeasures were systematically imposed, and slot machines were "easily circumvented by the average juvenile."[27] As in many other cases, push-button designs could succeed and simultaneously fail because of the simplicity of their design, paving the way for exploitation among those with a little ingenuity. The issue of honesty in vending transactions played a critical role in the development of the mechanisms at work, and early investors were forced to catalog the various ways one could fool a coin-operated machine (one noted an astonishing 38 ways a particular machine could be defeated).[28] The more buttons were touted as accessible and meant for anyone, the more that category of "anyone" caused problems.

As inventors developed the vending machine concept, they promised that consumers should have complete control over the products they purchased without interference from a clerk. Thus, it was necessary to ensure that this control adhered to designers' desired ends. Issues of control and automaticity figured importantly into both the design and promotion of these devices. An article about a cigar vending machine noted that, "under the control of the customer by means of a push button," the dispenser would "automatically" put an unscathed cigar of the purchaser's choice in hand's reach.[29] Similarly, the *Washington Post* touted a new machine that would allow a patron to "press the button and be your own milkmaid" as it dispensed small portions of milk for those who couldn't afford large supplies.[30] Vending machine designs capitalized on simple control (with machines handling the brunt of the work) as their main appeal—one could procure a cigar or take on the role of a milkmaid—through the swift effects provided by a push button while keeping the mechanism out of the way and at a distance from the user's concern.

Like other push-button novelties, vending machines used simple labeling of their push-button mechanisms to contribute to an idea of gratifying, automatic consumption. Next to large push buttons on typical vending machines, lettering would inform the customer to "PUSH," as in the case of the "Jolly Fellow" Tutti Frutti machine that distributed gum for one cent (see figure 5.4).[31] Similarly, a Fleer's gum and Chiclets dispenser stated plainly, "Deposit one cent then press the plunger." Beneath the text, a row of four rods featured the caption "Push."[32] In some cases, buttons were not pushed but pulled, hearkening back to a common "pull" model of the mid-nineteenth century; a nickel-in-the-slot machine, for example, instructed its user to "Drop nickel in slot/Pull button."[33] Instructions that accompanied buttons in certain contexts (such as hotels) might ironically cause confusion or lead to miscommunication. Vending machine signage, however, aimed to facilitate simple purchases that did not require signaling or human interaction with store employees. Users negotiated this transition with varying degrees of success and enthusiasm.

Buttons acted as go-betweens for consumer and machine, requiring but a small bit of effort from the presser before delegating control to the machine. Indeed, according to a *Scientific American* review of a new design for a pharmaceutical vending machine, "This automatic dispenser of course makes no mistakes."[34] Yet the author of the piece worried about potential backfires from such automaticity: "If the customer accidentally presses the wrong button, he alone is responsible for the error. Is this really what we are coming to?"[35] Where the self-service model could theoretically do away with unnecessary labor and tie pleasure to a push, it could also raise questions about the button pusher's role and responsibilities as a digital commander when caught between a hands-on and hands-off service approach. Pushing the wrong vending button would not blow up the world, but it could put undue pressure on the pusher by adding higher stakes to the transaction.

Figure 5.4
Vending machines employed the simple push-button mechanism to make consumption effortless and instantly gratifying.
Source: "The Jolly Fellow All-Iron Vending Machine," *The Voice of the Retail Druggist* (November 1910): 487. Image courtesy of the Warshaw Collection of Business Americana—Vending, Archives Center, National Museum of American History, Smithsonian Institution.

Such critiques of automaticity referenced these feelings of user uncertainty. For scholar George Herbert Palmer (1903), this experience made him feel not more powerful but rather that he played but a limited role in pushing a button: "The pressure of my finger ends my act, which is then taken up and carried forward by automatic and mechanical adjustments requiring neither supervision nor consciousness on my part. ... My finger tips, my lips, my nodding head are the points where I part with full control, though indefinitely beyond these I can forecast changes which the automatic agencies, once set astir, will induce."[36] This account of finger pressure as the end of an interaction, rather than the beginning, demonstrates why uneasiness often rose to the fore in discussions about push buttons. If indeed "automatic agencies" did take over, then button pushers played but a minimal role in the process. A presser could, after pushing the button, "leave that relatively unknown territory between the end of his finger and the delivered mechanical motion" to the machine and the electrical expert—according to another writer.[37] This design feature of buttons—obfuscating what happens behind them—led one historian to comment in 2014, "All these controls suggest to their user that the pressure of a finger can influence the course of events. But the invisibility of the actual mechanism confounds any confirmation of this hypothesis."[38] This perception of limited agency—and confusion over how buttons and fingers worked together to bring about a particular effect—made it all the more difficult for button pushers to take responsibility for their pushes or see themselves as moral, ethical agents. Perhaps more important than the question of agency, if a button pusher could stimulate dramatic effects in bodies and machines alike (regardless of whether she truly caused those effects), then various sectors of society had to think carefully about who should have access to buttons in the first place.

As with consumer technologies such as vending machines and elevators, a push-button mechanism on consumer cameras also attracted to the field of photography a new set of users who desired a more automatic and self-service experience without any requisite skill set. Through a catchy and direct tagline, "You press the button, we do the rest," Kodak cameras appealed to potential consumers by heralding buttons as beacons of simplicity and automaticity—just press the button and let someone else deal with what happened next (see figure 5.5).[39] The slogan served as one of the most popular anthems of the time period. One author noted that the phrase "is heard on the street, in the cars, at the theatre, in the clubs, and, in fact, wherever men and women most do congregate. The comic papers have burlesqued it, statesmen have paraphrased it, and it is repeatedly used to

THE KODAK CAMERA.

" You press the button,

we do the rest."

The only camera that anybody can use without instructions. Send for the Primer, free.

The Kodak is for sale by all Photo Stock Dealers.

Price, $25.00—Loaded for one hundred pictures.

THE EASTMAN DRY PLATE AND FILM CO., ROCHESTER, N. Y.

Figure 5.5
A successful Kodak camera slogan from the Eastman Company celebrated a photographic process that granted simple control to amateurs.
Source: Eastman Company, *Outing* 15 (1890): 12. Image courtesy of Harvard College Library via Google Books.

point a moral or adorn a tale."[40] Another editorialist called the statement the "prophetic cry of the age" because it spoke to one's ability to summon anything—and anyone—with a touch.[41] Recognized as one of the most effective advertising campaigns of its time, an advertising monthly journal interviewed George Eastman to understand how the phrase achieved so much success that "it has come to be common property, and is applied in every-day conversation everywhere."[42] In the article, Eastman noted that he never imagined how widely the phrase would take hold and could only offer in the way of explanation that he modeled it after the idea that, most basically, pressing the button represented the only work required of the camera's user. The popularity of Eastman Kodak's phrase in a wide variety of contexts indicates how the button served as a symbol of consumption, and pleasurable control of use for its users.[43] As a catchphrase, "You press the button" tried to remove all previous assumptions that consumers might have about photography, refiguring the enterprise as a hobby made for amateurs who need not fear the camera, its mechanisms, or the effort involved in producing or developing a photograph. By removing labor from the photography process and by putting the working parts of the camera at a distance from the photographer, Kodak offered a model of digital command that relied on pushing and effects.

Despite the slogan's clear success, this push-the-button mantra generated an irate response from those who felt threatened by the simple mechanism's invasion of a once skilled craft, invoking concerns similar to those that manifested in discussions about doing away with elevator operators, shop clerks, or factory workers and in debates about push-button managers as nonworkers. In reaction to the campaign, a large ad in *McClure's* magazine read (1896), "Don't Be 'A Button Presser,' "for he is a poor specimen of a photographer who is content to press the button, let others 'do the rest,' and then claim the results as his own" (see figure 5.6).[44] The stinging reproach, taken out by *Photographic Times* magazine, appeared in hundreds of publications at the end of the nineteenth century and criticized a growing number of amateurs using Kodak cameras. Boiling over into a variety of editorials and how-to manuals about photography,

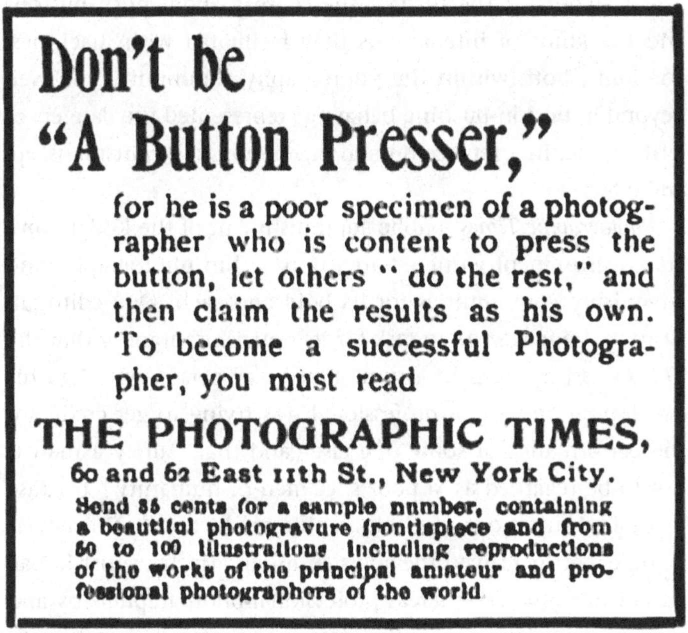

Don't be "A Button Presser,"

for he is a poor specimen of a photographer who is content to press the button, let others " do the rest," and then claim the results as his own. To become a successful Photographer, you must read

THE PHOTOGRAPHIC TIMES,

60 and 62 East 11th St., New York City.

Send 35 cents for a sample number, containing a beautiful photogravure frontispiece and from 60 to 100 illustrations including reproductions of the works of the principal amateur and professional photographers of the world.

Figure 5.6
This riff on the Kodak slogan represented a widespread anxiety about developers' new role in the photography industry.
Source: *Photographic Times*, "Don't Be 'A Button Presser,'" *McClure's Magazine* (1897): 40. Image courtesy of University of Michigan Library via Google Books.

this tirade against button pressers served as a direct affront to the Eastman Company's product and associated tagline. Professional photographers feared the rise of "you-push-the-button automatons" in their profession who would replace photo developers and make skilled workers unnecessary.[45] On one level, *Photographic Times'* antagonism toward the Kodak camera and its button helped to promote the magazine on the coattails of a well-known gimmick; on another level, however, it expressed a real anxiety at the turn of the century about push buttons and the kinds of interactions they facilitated with machines. For some, both within the photography community and well beyond it, button-pushing behaviors represented the dangers of letting machines replace human ingenuity, skill, craftsmanship, and effort.

Photographic Times' public admonishment of the Kodak camera advertisement garnered attention within photography and advertising communities for its bold approach. One editorialist took the magazine to task for its criticisms, arguing that the *Times* "had no right to assume that every person who had his developing done by a professional was trying to get credit for the performance of some one else" and that "surely a man is not to be regarded as 'a poor specimen of humanity'" because he desired this mode of photographic production.[46] This advertising expert perceived the ad as an unfair attack against Kodak hand camera users, whereas professional photographers by and large embraced the notion that pressing a button had nothing to do with the art of taking or developing a photograph. Photographers cast aspersions against button pushers along a number of fronts, deeming them "careless, slovenly" individuals who could not be expected to imbue a photograph with "the emotions of a man's soul."[47] These insults equated buttons with laziness and the kind of thoughtless behavior that existed in direct opposition to art making. The hand camera and its button threatened these individuals' livelihoods, opening the doors of photography to a new class of amateurs previously shut out from the expensive, time-consuming process of developing pictures. Although unsurprising that this rank of consumers would attract vehement reproaches on the part of work-a-day photographers and developers, it is more interesting to note the ways that buttons grew entangled in such a debate. Photography experts did not blame the camera, but rather its button, for encouraging automatic (and therefore dangerously simple) photography. The button came to serve as an emblem of the camera, its push representing the opposite of photographers' values and ethos.

At various points at the end of the nineteenth century, voices cropped up in support of button pressers, and as consumers increasingly took up the Kodak, photographers and developers were forced to come to terms with a change in the industry. In a stirring editorial titled "A Plea for Button-Pressers," one author wrote, "When all is said and done, the sun takes the photograph, not we, and whether we put in a holder, draw a slide, remove the cap, count four, or only press the button, we do not ourselves really and truly displace the silver particles on the gelatin. ... Why draw a hard and fast line between drones and busy bees at the mere development of a plate?"[48] Once again, questions of effort and authentic engagement stood at the heart of this debate: Did pushing a button diminish one's accomplishment or, as the writer suggested, merely combine steps in a process not entirely carried out by the photographer in the first place? Masking and even eliminating parts of the photographic enterprise, push buttons catalyzed concerns over whether simplicity stood in opposition to quality and true handicraft.

Although debates over photography continued, it is difficult to overstate the widespread influence of the "You press the button" slogan on other industries—and on the public's impression of push buttons generally, well into the twentieth century. A brief sampling of advertising slogans that riffed on the tagline demonstrates its potency. Drawing on the idea of putting the burden of effort on the machine rather than the button pusher, the *Chicago Daily Tribune* tipped its hat to the advertisement with regard to vending machines, suggesting, "The customer, for example, will drop his coin, turn the pointer to indicate the particular kind of goods he wishes, and touch the button. The machine will do the rest."[49] The American Supply Company of St. Paul, Minnesota, similarly offered a popular push-button novelty for 15 cents, as described in chapter 4, promising, "You press the button—the button will do the rest. Expose the button to your friend, he will be sure to push it and get a shock never to be forgotten" (see figure 5.7).[50] Riffing on everyday encounters that individuals might already have had with push buttons, this practical joke perverted the ordinary nature of buttons by making them unfamiliar and even dangerous under the guise of fun; the "rest"—the shock—remained concealed behind an opaque and unassuming button. It also took advantage of the fact that, in general, people enjoyed pressing buttons and often couldn't resist them; this psychological "weakness" lured in participants.

The Watrous Co. of Chicago also ran an advertisement proclaiming, "Push the button and we do the rest: PUSH and our No. 7 Button Faucet will deliver the water instantly, STOP PUSHING and the flow of water ceases. You can't leave *the water* running" (italics original).[51] This ad emphasized the button pusher's full control, where pushing started and stopped the action at

Figure 5.7
Electric button novelties that shocked their pressers were popular in the late nineteenth and early twentieth centuries as part of a culture that valorized tomfoolery, painful pranks, and electrical experimentation. Source: *Popular Mechanics* 8, no. 3 (1906): 382. Image courtesy of Google Books.

will. In other cases, the "rest" involved being served by a human attendant, as in the case of a prototype "drive thru" grocery store. According to the store's description:

Drive up in front of our Grocery Department. You will see a new box post standing on the edge of the pavement with an Electric Push Button. Take your whip handle and press the button and a real live clerk will rush out at once, anxious to take your order, and to save you the trouble of getting out, hitching your horse and entering the store. Try it and see how it works. This principle, "You Push the Button and we'll do the rest," you will find now obtains in every department of our business.[52]

It is noteworthy that the slogan often applied to both human and technical help; it mattered less whom or what carried out the button's task and mattered more that the task would be completed "at once" by an attendant that remained out of view until needed.

The Royal Easy Chairs company offered a playful spin on the popular phrase, with copious advertisements for its reclining chairs that allowed the person sitting to "'Push The Button—and

Rest," putting buttons and comfort always within an arm's length (see figure 5.8).[53]

The ad combined sedentarism and button pushing to portray the luxury of digital command and the privileged position of the relaxed button pusher.

Beyond advertising, Kodak's catchphrase came to signify a common conception of buttons as facilitating automatic (and thus conversely undesirable) experiences. Where women's "electric buttons" were previously analogized in reference to bells—"jangling" and ringing discordantly—the Kodak analogy came to serve as a new paradigm for understanding the supposedly automatic and therefore largely uncontrollable response of a woman's body. In the words of one commentator reacting to Dr. Morris's analogy of women's clitorises as electric buttons in *Western Medical Reporter*, "That's right! In the language of the Kodakers, when a man 'presses the button' the woman 'does the rest.'" Men could activate a sexual response in women's bodies like a basic consumer camera—setting their mechanisms into motion at an instant with a push.[54] Thinking about women's sexual organs as buttons supported a medical ideology of subduing the female body and remedying it from a supposed disorder of control—one had to begin by going to the "electric center" of the woman, which "normally sets the whole mechanism in operation"—the button.[55]

Although Kodak's anthem served as a fitting metaphor for physicians trying to make a case for treating women's bodies in a particular way, laborers applied the slogan to express fear over automaticity in workplaces. Across industries, workers feared that button-pushing practices would create a new rank that would make them unnecessary and also steer the course of human labor in an irreparable direction. One writer for the

Figure 5.8

A luxurious present imagined by Royal Easy Chairs, which promised users they could control their relaxation at the push of a button.

Source: Royal Easy Chairs, "Push the Button—and Rest," *McClure's Magazine* 44 (1915): 142. Image courtesy of Google Books.

Railroad Trainmen's Journal drew on Kodak's tagline in an assessment of the state of affairs for laborers: "We are living in an era where 'you push the button and we do the rest' can be said of many devices and where the button accomplishes the task the man loses the wages, and the 'button' does assist him to lose his means of living."[56] The author deterministically suggested that buttons possessed the power to ruin a man, taking over the work of a human hand and stripping that hand of its wages. Another similarly surmised, "All that many a workman needs now is an ear to hear an order, an eye to see an electric button, and a finger to touch it. Electricity and the inventor do the rest. Dependence upon machinery is becoming almost an instinct."[57] Distinguishing working bodies from button-pushing bodies, this opinion reflected the perspective that hands' roles were now reduced to a finger. The working person could rely on the slightest and individual parts of his body to carry out his work. In this regard, the act of pushing buttons signified a fundamental shift in bodily ways of working and being a worker that emphasized digitality or finger operation.

Workers—much like photographers—imagined button-pushing practices as a potential source of ruin that would replace skilled workers with unskilled ones, leaving behind atrophied and incapable hands. One craftsperson worried to this effect, "We have not reached the point yet where the majority of us do nothing but push buttons, but we are fast approaching it. What then? ... You think a republic could stand that was composed of nine-tenths of button-pushers and the other tenth of captains of industry?"[58] This notion that everyone would perform the functions of either a manager or an untrained button-pusher penetrated numerous fields. A "push-button habit" caused concern that such stratification would leave a gaping hole in industry where no one would carry out "actual" work.[59] Of buttons' many critics, socialists most vehemently advanced this line of attack by scapegoating push buttons, holding them up as a metaphor for capitalism's ills. In a treatise on socialism, for example, author Charles Henry Vail (1899) worried that "the time will come when the work of the world will be accomplished by simply pressing an electric button"; who would own this button and control the workers behind it produced great concern. "When a few, by simply pressing a button, produce the goods, then the great multitudes will be unemployed," suggested Vail, "their consumptive power gone, and they themselves reduced to degradation and starvation."[60] In these cases, buttons served as compelling icons for fears of consolidated, private ownership over machines where the rich got richer and the poor were removed from the equation entirely. Importantly, Vail, Benson, and others often thought in future terms. Realistically, buttons could not materially accomplish every kind of laborers' task at the turn of the twentieth century, but the fear that eventually they could weighed heavily on observers.

To respond to increasing mechanization and automaticity, thought leaders in manual or formerly manual industries focused on the importance of human contact, human intuition, and human skill to effectively carry out the operations of a business. One author proposed a compromise in recognizing that although push buttons would continue to exist, workers would simultaneously need to exist to create and maintain the buttons; through these efforts, they could achieve mastery over machines.[61] To this end, educators in the field of industrial arts urged that machine workers find a way to always maintain agency in the growing "battle" between humans and machines. Author S. N. D. North made remarks to this effect in an 1896 commencement speech at the Pennsylvania Museum and School of Industrial Art: "I plead ... for the education of the men and women who control and manage the machine, not merely that they may minister to a higher public taste, but that they may save themselves from the degradation of slavery to the machine."[62]

Similarly, Charles H. Ham (1900) wrote in a book on the relationship between mind and hand that man would only thrive in a machine age if he could "harness" electricity and steam for his purposes.[63] Ham's vision of manual labor—of hands and their touch—sought to preserve corporeality at a moment of destabilization in the category of "hands" and what they could do. He exalted human bodies but did not see them as incongruous with machines. In defense of this view, Ham quoted Dr. George Wilson, who suggested, "In many respects the organ of touch, as embodied in the hand, is the most wonderful of the senses. The organs of the other senses are passive; the organ of touch alone is active." According to Wilson, "The hand selects what it shall touch, and touches what it pleases. It puts away from it the things which it hates, and beckons towards it the things which it desires."[64] In his defense of touch as an "active" sense, the doctor imagined hands as delegates or agents for the body as a whole. Yet the act of pushing a button continued to be viewed as the antithesis of human touch, manual labor, or active engagement. So argued William L. Price in a speech to the Eastern Manual Training Association in 1903, who warned his constituents that "we have gone machine-mad" in an effort to do work that would require only pushing buttons.[65] Price positioned the act of pushing buttons as inhuman in its detachment. This commentary and others popular in the machine age commonly focused on how humans could dominate over and sublimate machines rather than being dominated by them, thus maintaining humans' importance—and hands' relevance—in the face of machine-assisted labor.

Both in the world of work and in the practices of consumption, people began to view buttons in terms quite different than before. Unlike the bell that would stop ringing once the finger released it, the "automatic" button suggested a new relationship between hands and machines predicated on delegating effort as well as responsibility. Where companies such as Kodak and Otis sought to sell this concept as an empowering one, freeing human operators from the need to strain, train, or even think, these attributes provoked fear and outrage among laborers of various kinds as they upended traditional definitions of "manual" and "craft" if anyone who pushed a button could achieve the same result as anyone else.

6 Let There Be Light

The comfort of never having to grope in a dark closet or to take a greasy candle in among one's clothes is one of the greatest comforts of a properly lighted house.[1]

Buttons designed for automatic consumer convenience—like those used for vending machines, elevators, and cameras, were constructed with the elements of digital command in mind. They provided button pushers a constrained set of choices to make the scope of decision making limited and effortless: push to start and stop, push to dispense, push to go up or down, and so on. This principle particularly took root in relation to lighting, where switches, especially in residences, took the shape of a push button. Interestingly, even before electric lighting, black and white button mechanisms, sometimes referred to as "press-buttons," "push keys," "pushes," or "automatic keys," functioned as electric switches (or burners) for igniting gas lights.[2] Their binary design—white for "on" and black for "off"—also capitalized on yet another binary opposition of light and dark. They worked by containing a vibrating magnet that would gradually turn on the gas, thereby delivering a series of sparks for the length of the closed circuit.[3] Proponents of gas lighting stressed that these buttons could provide "instant" light without the need to fumble with complicated parts, trigger gas on demand, and ensure the utmost safety.[4] When it came to controlling these lights, one need not twist or turn one's hand to titrate gas; rather, pushes would require a form of hand engagement only at the beginning and end of the process.

Still, as electric light mechanisms improved, growing concerns mounted about the safety, efficacy, and cost of gas as a solution for lighting.[5] Some complained that the "push did not always do what was expected of it," and the "fickle" and "delicate" apparatus caused trouble to gas consumers when electrical connections failed. Sometimes, users could turn gas on but not off, or they could turn it on but it would not light.[6] Gas and electricity industries vied over the "better" source of power for a number of years, with each marshaling resources, allies, and technologies to defend their energy source.[7] While gas-powered lights continued to exist well into the twentieth century, electric lights with push-button switches increasingly appeared as desirable and practical options that would displace gas methods.[8] As part of this transition, early adopters of lighting confronted differences (or a lack thereof) between gas keys and push-button switches. According to prominent electrician Frank H. Stewart (1913), button pressers in earlier times often confused gas light buttons from electric light buttons since both employed one light- and one dark-colored push (see figure 6.1).[9] This similarity meant difficulty, for some, in adapting to wholly electrical appliances, but it also introduced push buttons as familiar points of contact that helped to bridge the gap between gas and electric power.[10] Indeed, if a consumer had been using gas and wanted to transition to electricity, an electrician could even mount gas keys and push buttons on the same switch plate, making room for the new alongside the old.[11] This design solution reflects the

Figure 6.1
Gas or "push" keys typically featured one light- and one dark-colored button to operate gas-powered lights. Switches for electric lights used the same design, often causing confusion between gas and electric service.
Source: Image courtesy of the Warshaw Collection of Business Americana—Electricity, Archives Center, National Museum of American History, Smithsonian Institution.

fact that, across a range of technologies that includes buttons, engineers throughout history have generally tended to import solutions from one infrastructure to the next.[12]

Given the binary way in which buttons worked, advertisers also reinforced the notion that users could control a switch by understanding its two conditions. Writing about the award-winning "Single Push Button Flush Switch" created by Woltmann & Trigg, the Electrical Specialty Company spoke effusively of its "simple and wonderful" mechanism: "That the mere pressure of a button will enable a person to turn 'On' or 'Off' the electric lights of an establishment, means a good deal more than these few words convey," emphasizing "mere pressure" as a novel facet of interactivity.[13,14] The widespread design choice of lighting via push button meant that switches could not "occupy any intermediate position" and did not allow for "tuning of the eye" to make gradual adjustments to brightness or shadows, as they had in the past.[15] In other words, the black and white buttons translated to literal darkness or light without any analog degree of control in between. As a result, a shift to on/off carried with it a particular—and significantly different—way of seeing and touching with physical, social, and political consequences. Electric light users were not only forced to adapt to the conditions of seeing or not seeing—they also had to make choices about whether or not to touch.

Black and white buttons—and the ON and OFF mechanism they represented—served as a potent cultural symbol. Indeed, metaphors regarding switches often extended to comparisons of life and death as well as to the electric death penalty, as in one piece of fiction wherein the protagonist mused about a near-death experience: "In an instant the full current of life, with all its unfulfilled purposes, and ties of love and affection, would have been brought to a stop. But I myself would have felt as little as an electric lamp when the current is switched off. The light would have gone out, but there would have been no pain."[16] A similar reflection in a medical journal upon the death of a physician noted that "in the midst of life there is death and no man knoweth when his hour shall come—it is but a press of the button and the lights are out."[17] This perceived toggling between current switched on or off, light or dark, and life or death could seem crude and unfeeling—as in the case of the electric death penalty described in chapter 4, but it also contributed to making death by electricity appear sterile, scientific, and civilized. Such comparisons also drew attention to the binary nature of ON and

OFF switches that put the electrical mechanism at a distance from the user.

The lighting industry particularly seized upon this binary relationship and the concept of mere touch as a way of selling light as safe, simple, and effortless to operate. "The perfect light must not be difficult to control," surmised the inventor and entrepreneur Robert Hammond (1884), using language similar to that of elevator manufacturers. "It has been stated that if the use of the electric light were general, our cooks, housemaids, and footmen would have to be trained electricians and mechanics in order to manage it. Well," Hammond demonstrated, "if you will watch me going the round of this room, you will see how unscientific are the means used to turn on and off the lights."[18] He referred to two specific features of these electric light switches—push buttons—near the turn of the twentieth century that closely corresponded with the tenets of digital command: First, laypersons should have easy and effortless control over electricity and need not possess special skills. Second, and interrelated, these switches should operate with the simplicity of binary (mutually exclusive) conditions; that is, one push for "on" and another for "off." While two-button switches were most prevalent, some designs sought to streamline the binary process even further, such as the "O'Brien Push Button" that only required one button for make-and-break.[19]

Despite advertising pronouncements, push-button switches, like their earlier bell-ringing counterparts—earned a reputation for their inconsistency in working properly from an early stage. Some complained—once again in reference to Kodak—that no matter how "beautiful it was in principle" to "push an electric button and the apparatus does the rest," a mechanism "so fickle could not be recommended" to consumers.[20] According to the electrician Granville E. Palmer (1911), one could hardly tell the difference between "good" and "bad" switches from their outward appearance, and when buttons failed, their users often blamed electricity in general rather than buttons specifically.[21] Manufacturers of buttons typically fell into one of two categories, the first of which created "a reputation for thoroughness that makes his product a standard throughout the world," and the second "equally well known" for "produc[ing] fair goods at a ridiculously low price."[22] The latter category of cheaply made push buttons often suffered from defects that included poor insulation, shoddy mechanisms, or corrosion due to dampness.[23] Constructing a high-quality push button required both effective materials and skilled laborers. One journalist, on the hunt to better understand how manufacturers produced buttons, visited the highly reputable Cutler-Hammer facility in Milwaukee, Wisconsin, and reported back on the complex process, describing the button's spring mechanism:

The C-H push-button contains very tiny but very essential springs and each spring must be bent and caught around in a circle. This is a three machine process. The wire is automatically wound, cramped and shot forward. As it touches the opposite wall of a gap, it creates a circuit which sends a knife through it on the instant; thus it automatically measures and cuts itself at the rate of 13,000 springs per day, none of which deviates more than a quarter turn in length. It then goes to another machine which thrusts up the last turns on each end of the length of spring. At the third machine, a deft-fingered operator catches the ends together and the circle of spring is complete.[24]

Although this look inside the multistep process of constructing a push-button switch might encourage some homeowners to invest in high-quality buttons, the fact that all buttons looked the same on the outside made it difficult not only to distinguish between buttons but also to determine whether one could blame a faulty button for electrical problems in the first place.

The first step toward making buttons act as the safe and tangible "faces" of electricity involved covering up the wires that made them work, a project similarly undertaken to encourage the use of electric bells in homes. Wiremen often conducted "wiring campaigns" to convince homeowners of the safety, cost effectiveness, and even necessity of outfitting their homes with electricity.[25] Because electricians were wiring many homes for the first time—or, in later years, replacing outdated wiring—they thought carefully about how individuals should interact with and control their lighting and its associated push buttons.[26] In order to lay wire properly and protect users from danger, builders typically took pains to "know where every electric button is to be" from the outset.[27] Still, with a lack of straightforward regulations or official inspections that applied to the whole electrical industry, some worried about accidents caused by electric current that could injure both electricians and "ordinary" citizens.[28] Electricians also complained about contradictory rules and codes, citing hurried wiring jobs and "American carelessness" as sources of concern.[29] In fact, no uniform practices for the electrical wiring trade would be implemented until well into the twentieth century; as a result, a great deal of variety existed in terms of wiring placement. In general, though, a shift occurred toward insulating electrical wires and tucking them away when previously no such measures were taken.[30] Wiring campaigns focused on educating homeowners about the ways that an electrician could "fish" wires through ceilings and partitions so that a change to electricity would "divest it of all its former terrors."[31] Electricians often faced an uphill battle in doing away with homeowners' prejudices, and concealed wiring offered one argument for electricity as a benefit rather than a nuisance.[32] Other appeals promised minimal disturbance to a house's existing structures, assuring homeowners that little dirt, disorganization, or disruption would result from introducing electricity.[33]

As with concealed wiring, electricians lauded buttons' design because they could blend them into their surroundings as "flush" or "sunken" switches that made everything but the button invisible, much as they could be attached to desks for push-button managers or affixed to tables for housewives (see figure 6.2). They viewed this flush design as "modern" and desirable for consumer use.[34] Architects were encouraged to purchase these button models because they could install them completely flat against plaster walls.[35] Entreaties made to consumers took a similar tack, with advertisements for buttons commonly referring to the flush nature of buttons. They noted that "nothing is presented to view but the face-plate, which may be so ornamented as to correspond with its surroundings."[36] Switch plates, therefore, likewise received attention from decorators who wanted to incorporate them "into the decorative scheme" and blend them in with the woodwork.[37] Particularly at the entrance to a home, a button should have an "approved" finish such as bronze, porcelain, or brass so that they should be "in harmony with the rest of the house."[38] In her recommendations to contractors wiring homes for electricity, Elizabeth Whipple declared that "the right kind of switches are always push button types and if you really want to make a hit with women, have switches plated to match the hardware."[39] A widespread viewpoint emphasized that sunken switches were "very much prettier than the ordinary protruding ones," once again reinforcing the notion that electricity should not "stick out" unnecessarily, a key difference between push-button designs and other kinds of switches.[40] To this end, buttons also decreased in size—no longer four or five inches high and three inches wide—and were often replaced by

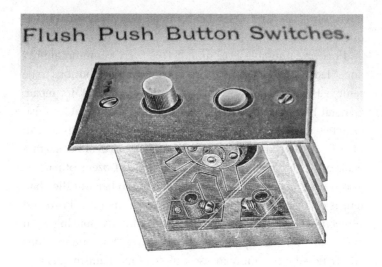

Figure 6.2
Electricians preferred switches of a "flush" nature that would blend into their surrounding walls. This style was viewed as the most aesthetically pleasing.
Source: General Electric, *G.E. Specialties Catalog* (March 1910): 149. Image courtesy of the Warshaw Collection of Business Americana—Electricity, Archives Center, National Museum of American History, Smithsonian Institution.

buttons called "midgets," which were known for their unobtrusive appearance.[41]

These "flush" buttons that protruded only slightly from their surroundings and blended seamlessly into the décor, a key facet of making the concealment of digital command possible, constituted more of an ideal than a practical reality. In fact, one of the primary complaints lodged against buttons was that their "sunken" nature made it difficult to maintain or replace them.[42] During installation, flush buttons often wreaked

havoc on plaster and wallpaper.[43] In addition, dust and dirt frequently accumulated inside buttons, causing them to jam up
and become unusable over time.[44] This "very common annoyance" led to new designs that could allow for push buttons to be
removed from the surfaces in which they were sunk for repair.
General Electric (G.E.), for example, created "removable" push
switches so electricians could insert buttons into walls in two
phases—during construction and afterward—to avoid damaging
walls or trapping debris inside the buttons.[45] Dozens of patents
also sought to address this concern, trying to balance the challenges of making buttons aesthetically pleasing (i.e., flush) and
protecting them from dirt and moisture while also making them
extractable for subsequent repair or change.[46] As one inventor
wisely noted, "To simply incase a piece of mechanism ... is not,
as a general thing, a difficult matter, but it is not so easy to incase
such parts and still have the electrical connections thereto readily accessible."[47] Concealment and accessibility therefore stood
at odds with one another, making buttons one of the least desirable control mechanisms for electricity.

Despite these noteworthy impracticalities, ideals of accessibility and simplicity merged together in thoughts about managing light. Just as homeowners might want a button in close
proximity for summoning a servant, electricians proposed that
they might also want controls for light close at hand. Electricians advocated that homeowners should mount push-button
switches approximately four or five feet above the floor for easy
pressing. To arrange buttons in the most effective and practical manner possible, electric wiring specialists and architects
devoted much attention to thinking about and planning how
homeowners should interface with buttons. They primarily
focused on efficiency, economy of movement, and the kinds

of communication and control that would occur in different spaces. Electricians recommended the installation of push buttons in routinely used areas, including near the head of the bed, in the library or sitting room, at the tops and bottoms of stairways, and at each entrance to the house (front, side, and rear doors).[48] Although experts often provided guidance on where buttons should go, homeowners took a stake in their placement, too, and could even "act out" how they might use buttons in different situations; for example, they could map out how their hands might reach in the dark to illuminate a room by pushing a button. To this end, F. C. Allsop recommended in 1894 that, when preparing to install push buttons, the homeowner should use the following method at night:

The best position is then found by putting out the hand, when in bed ... towards the wall and making a mark where it touches. This is done several times, changing the position once or twice, and making a mark each time the hand touches the wall. In the morning, a point centrally of all the marks is then taken and the push fixed at this point, when it will be found that in future the hand, when put out in the dark to find the push, nine times out of ten comes at once upon it.[49]

This suggestion well preceded a formal human factors movement, but it recommended practices common to that field by suggesting that homeowners could simulate the push-button gesture repeatedly to devise the most ergonomic and intuitive arrangement that would minimize bodily effort. In general, according to one homeowner, "When you are dealing with switches it isn't a game of 'button, button, who's got the button' if you really want to avoid inconvenience in finding your switches. They should be placed where strangers, even, can quickly find them."[50]

Yet the problem of putting buttons "within reach" arose once more, as it did in so many spheres of life. For example, housewives familiar with early forms of electricity often warned that push-button switches in young boys' hands would lead only to troublemaking; thus, they discouraged youths' experimentation with and access to buttons. In her 1891 book *Decorative Electricity*, J. E. H. Gordon, an electrical engineer's wife, wrote, "The switches should be placed high in the nursery and schoolroom, and strict rules should be made that the children do not meddle with them, as they will climb on chairs and footstools, and electric switches, as I have found from personal experience, are perfectly irresistible to little boys."[51] Although children were often encouraged to learn about buttons and incorporate them into their understanding of the world, many believed that the psychological appeal of the switch would prove too great to resist for a young hand. Those with a bit of forethought therefore planned to make control inaccessible at the outset of wiring homes, buildings, and other devices. Just as electricians might cover up wires so they would remain out of sight, wiring manuals suggested that electricians should install switches between 3 feet 10 inches and 4 feet above the floor to be "out of reach of small persons (children), who would perhaps make toys out of them."[52] Do-it-yourself articles often included suggestions to mitigate inappropriate touches, as in the case of automobile manufacturers that devised ways to lock up lighting switches from children fooling around.[53] They could guard against "many small boys [who] take pleasure in unscrewing covers from electric push buttons" and received attention in forums like *Popular Mechanics* magazine so that tinkerers and amateur electricians could institute precautionary measures inexpensively.[54]

Beyond keeping children in line, creating conditions for pushing buttons the "right" way involved managing buttons themselves and developing a set of protocols for appropriate touching in particular situations. In fact, as with efforts made by homeowners to manage how servants used electric bells and responded to calls, employers similarly limited how their staff could make use of light switches. Housekeepers often possessed a master switch in their bedrooms that would control the lights in younger maids' rooms in the morning and at night. Those servants who did have access to the switch might receive routine reminders to turn off the lights when leaving a room to keep costs down.[55] Homeowners might also install buttons with timers to manage when servants could push them, thereby avoiding situations in which servants might forget to properly manage lights or alarms or intentionally leave them turned off.[56] Household roles delegated who should have the right to push buttons and to what degree they were permitted to control lighting, creating a set of protocols around touchability.

These discussions often arose in regard to "waste" of electric current, with some worrying about "abnormal consumption" of current if electricians did not arrange circuits and push buttons properly. Others complained about an indiscriminate use of buttons leading to overspending.[57] At this time, artificial light, according to one historian on the subject, was "employed in a rational, economical way, not as a vehicle for conspicuous consumption."[58] Within this context, then, Gordon recommended that a lady of the house keep switches for decorative lights protected behind lock and key in a wooden cupboard (with appropriate ornamentation to match the room's décor), adding, "I consider that a great deal of electricity is wasted by people turning on the light to show their friends."[59] Similarly, one manager of an electric light company complained to this effect in an editorial about his wife's fixation with keeping light usage in check: "Her idea of an electric light is a devouring monster—something to turn off," he wrote to the *Rotarian*.[60] "When a light burns a moment longer than she thinks necessary, or if one light will do instead of two," he continued, "she is utterly miserable and cannot think of another thing till the offending light is turned off."[61] Although he found this attitude extreme—and recounted her neurosis in humorous detail—many homeowners expressed similar worries over keeping the lights on for too long.[62] These concerns led electricians and push-button manufacturers to devise various strategies to help manage consumers' electricity usage by placing technical constraints on buttons. Automatic door switches (especially for closets and basements), for example, would turn lights off upon closing the door in order to deal with "carelessness" and "annoyingly frequent occurrence[s]" of lights being left on.[63] By putting these "automatic" measures in place, one could discreetly manage fingers without needing to curb the urge to touch.

Interventions like the ones described reveal that although the electrical industry sold button pushing as uniformly desirable, many users took great pains to control ON and OFF switches and manage whose fingers should have access to these conditions. The question of push-button control extended well beyond the button itself, raising questions about who could touch, through what kind of finger action, and with what authority. To make buttons work the way that inventors imagined them to required learning what a wide range of users might do and then devising ways to circumvent those maneuvers. A wide chasm existed between idealized button pushers, who represented the notion that "anybody" could push a button, and the actual behaviors of children and adults whose mischief, laziness, overenthusiasm, or confusion necessitated countermeasures.

Needless to say, these conflicts never entered the discourse of advertising. Because consumers were so suspicious and uncertain about electricity—particularly the idea of it in their homes—appeals toward selling push-button electricity sought to do away with what electric companies perceived as awkward, inefficient, and even dangerous hand and body movements that threatened safety. A recurring trope in ads targeted at potential users promoted the use of buttons as a ward against darkness and a remedy for the unseeing body; these ads assured consumers that push buttons could bring safety and comfort within reach. One tactic employed by manufacturers and distributors of push buttons and their associated devices involved denigrating the groping gestures and stumbling common to those used to moving about their homes in darkness (pre-electricity).[64] Making a pitch to those accustomed to injury, G.E. asked potential consumers, "What of comfort is there in stumbling across a dark room until you manage to find a switch at the opposite door?"[65]

J. H. Bunnell & Company's promotion of its push-button wall switch featured a proclamation along these lines, for the button, which was engineered to light up in darkness, "enables a person to light the gas or ring a bell in the night, *at once*, without searching in the dark to find button or matches" (italics original).[66] Similarly, in a promotional storybook written in the early 1900s and titled *Solid Comfort, or the Matchless Man*, the Edison Electric Illuminating Company sold the push button as a savior to a domesticated man afflicted with dangerous bodily movements: "No stumbling over the furniture! No breaking of bric-a-brac! No black cats, coal piles, dark cellars and midnights! Not a trace of them! Light, light at a touch, and plenty of it!"[67] An illustration of a man with his finger upon the switch accompanied these words, emphasizing the powerful simplicity of an electrified touch that could control light from a distance (see figure 6.3).

The corporation also created an illustrated comic as an ode to electricity in 1906; here, again, the button fended off hazard.[68]

When to the furnace this man goes
He does not mutilate his toes;
He doesn't get a nasty fall;
He touches the button on the wall.

Figure 6.3
Depiction of a man using a push button, symbolizing modern interfaces triumphing over outdated modes of summoning light.
Source: Edison Electric Illuminating Company, "The Edison Man," 1906. Image courtesy of the Warshaw Collection of Business Americana—Electricity, Archives Center, National Museum of American History, Smithsonian Institution.

This depiction of "The Edison Man," although couched in a playful tone, portrayed a modern user, a digital commander, as one in control of his situation, body, and technology through the use of a push button. The Edison Company even created jingles about electric light, featuring buttons prominently in these irreverent odes. In one such jingle, touching a button caused the fictional user, Harold, to proclaim that day(light) "springs into being from the loins of night,/And makes my path clear with Electric Light./No smell, no match, nor fume to do me harm./I wouldn't be without it for a farm."[69] These catchy entreaties to consumers promised that homeowners—especially male homeowners—could access a safe and unfettered form of power that transformed once failing hands into finessed ones. Where women could sit back and relax, commanding and communicating with their household staff by pressing a button that would ring a bell or annunciator, men, too, could preside over their domestic environments—coming and going as they pleased at any hour of day or night—by controlling electric lights with ON and OFF switches.

Years later, in 1913, the Edison Company continued its promotion of button switches as modern triumphs of light over darkness and organized bodily movements over disorganized, dangerous ones. In the poem "Let Us Go Back," Roscoe Gilmore Scott wrote:

Let us go back to the candle-light
To those famous 'good old days'
To the good old dark, to the good old plight
Of a stumble in the haze:
Let us feel the fear that we used to know,
When the midnight fire-bell clanged;
Let us search for matches high and low,
That the thief may go unhanged!

Let us go back for a minute slight—
Then press the button and have real light![70]

Each of these examples identified the past—a life without button pressing—as a horror (injuries incurred, thieves running loose, fear and stumbling in darkness), and presented modern life as a salvation marked by convenience and comfort. Advertisers used the binary relationship of light and dark to associate darkness with danger and filth and light with brilliance and safety.[71] Domestic space no longer threatened with its potential for hazards or the uncertainty posed by its vastness; buttons could cut through that space to make it knowable and traversable, and button pushers could act as operators at a distance from the lights they sought to control. Promotional materials also emphasized that a button presser could physically grasp this seemingly intangible light, advertising "Light At Your Finger's End; Every minute of the day or night" and "Have All Your Machines At Your Finger Tip."[72] As with depictions of "mere touch," the fingertip yet again figured importantly as the site where hand control and digital command occurred.

The desire to make light "handy" and reachable crystallized in the form of flashlights (known during this time period as "pocket lights"). These lightweight devices, weighing approximately seven ounces and measuring barely an inch thick, were equipped with a push button on the side to provide instant light for circumstances including finding a house number at night, locating a dropped article, or avoiding dangerous places and dark streets (see figure 6.4).[73] Companies like Eveready suggested that the pocket light could aid men in numerous occupations—from miners, hunters, and plumbers to sailors, physicians, and policemen—in a fashion similar to that of the scarf pin light.[74] The transportable nature of these technologies meant that the

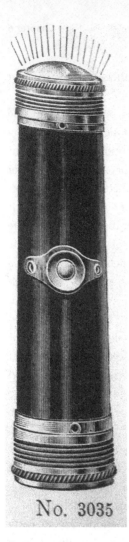

No. 3035

Figure 6.4
"Pocket lights," later called flashlights, transformed the concept of
the electric scarf pin light into a practical, portable technology used
especially by those with nighttime occupations.
Source: Adams-Morgan Company, *Electrical Apparatus and Supplies: Dynamos, Motors, Chemicals, Wireless Telegraph and Telephone Instruments,
Books and Tools*, ca. 1911. Image courtesy of the Collections of the Bakken Museum.

user possessed binary control without tethers to any particular location.

Whether through light in one's pocket or from a lamp, the electrical industry tried to encourage consumers to embrace electric lighting by highlighting the simplicity of controlling light with one's hands. Early users of electricity expressed wonder that, "at the present time when, by the mere touch of a finger on a button, we can instantly obtain a flood of brilliant light, it is difficult to realise what must have been the conditions of life when man had to be content with the smoky flame of the pine torch or of the rush dipped in olive oil which burned with but a feeble light even in its vase of finest workmanship."[75] Although light itself garnered significant attention, commonplace mentions of "mere touch" in reference to electricity did not aim to delight the eyes but rather emphasize slight touches, effort, and methods of control.[76] Companies like the Cutler-Hammer Manufacturing Company invoked the sparseness of hand effort in advertisements, asserting that only "one hand is needed [to operate lights]—three fingers to steady the socket and the thumb to press the button" (see figure 6.5).[77]

This emphasis on reachability began to take shape as "remote control" in the 1900s—an extension of earlier experiments with telegraphy and toys to stir machines into action from a distance. From a technical perspective, electricians considered how a touch in one place could cause an effect in another, particularly in response to consumer complaints about an "inability to be at all the places at one and the same time" to turn off switches.[78] Large-scale spaces, especially commercial ones, for example, required plans to control large blocks of light and other apparatuses from a distance. These switches could also prove useful for controlling circuits in "factories, hotels, apartments, public

Figure 6.5
"Mere touch" as illustrated by an advertisement for one-handed operation.
Source: Cutler-Hammer Manufacturing Company in *National Electrical Contractor* 16 (1916): 98. Image courtesy of the New York Public Library via Google Books.

buildings, stores, piers, and elsewhere wherever it is desirable or convenient to control any electric circuit from remote points."[79] They typically worked automatically, using a magnet to close and open the circuit and therefore requiring no human intervention at all, whereas semi-automatic circuits needed an operator to push a button that would spring the magnet into action. A "Remote Control Switch" made by Pettingell-Andrews—first installed in the New York Library in 1908—offered such a solution.[80] Through the "mere pressing" of push buttons, the user could "control any portion or all of the lights or power in a building by a push button located on the desk of the superintendent or any subordinate employee."[81] Another similar switch device allowed the operator to use a push button to maintain control "in any part of a building at any time."[82] Other early installations also occurred in locations such as the Prudential Life Insurance Company in Newark, New Jersey, and the New York State Education Building in Albany, New York (see figure 6.6).[83] Railroad operators similarly made use of long-distance push buttons, arguing that "centrally located distant control" could enable "the saving of many lives" in the event that a train needed to stop at a moment's notice.[84]

Electrical journals enthused about the ways that switchboards, which featured groups of push buttons, could allow a button pusher to achieve multiple effects at the desired time and in concert with one another, as in theater performances where remotely controlled buttons could open or close curtains and doors and turn lights on and off on demand; this form of control offered a powerful spectacle of invisibility while acting as a safety mechanism in case of fire or other emergency.[85] These apparatuses functioned not to send a message or signal across distance, such as pushing a button to ring a bell, but rather as mechanisms to effect change in the condition of another technology: alongside theater effects, the remote button could illuminate a light bulb, set a thermostat to a certain temperature, or immediately halt a train or equipment on a dangerous course. Such interventions gestured toward numerous possibilities for control at a distance. Electricians often referred to brass switch plates or "gangs" of multiple push buttons as a "'lazy-man' switch"

This Trade Mark The Guarantee of Excellence on Goods Electrical.

G-E Remote Control Switches in New York State Educational Building

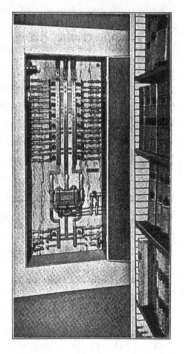

The illustration shows a G-E Remote Control Switch installed in the stack room of the Legislative and Law Reference Library in the new State Educational Building at Albany.

This switch opens and closes the lighting circuit for the stack room corridor, which is about 200 ft. long. The switch can be operated from any one of ten push button stations, located at the different entrances to the room. A person can turn on the lights when entering through one door and can turn them out when leaving by any other door.

A number of G-E Remote Control Switches control the lights in other stack rooms in this building.

G-E Remote Control Switches are convenient and reliable, and their operation consumes a negligible amount of power, as current flows in the coils only when the switch is opening or closing.

Our Bulletin A-4070 describes these switches in detail.

Figure 6.6

An advertisement detailed General Electric's remote-control switch in a reference library in Albany, New York, controlled by one of ten push-button stations located at entrances around the room.

Source: "G-E Remote Control Switches in New York State Educational Building," *Journal of Electricity, Power and Gas* 33, no. 3 (1914): 12. Image courtesy of University of California, Davis Library via Google Books.

to emphasize the ease of pushing buttons centrally located to achieve effects in multiple locations.[86]

Push-button light switches were prominent in the United States in the late nineteenth and early twentieth centuries, but for the most part, the rest of the world adopted other kinds of switches instead. According to a writer in the *London Electrical Times* (1912), "Another thing I could not get explained [in America] was the general use of push-button switches. These are not nearly so convenient as tumbler switches, and must be a lot more expensive."[87] International observers repeated this sentiment across a number of years in their remarks about selling American electrical goods abroad. "The push pattern finds considerable favour, in America," a British author wrote in 1918, and in 1919 a French importer suggested that "push-button switches of the American type would be regarded as luxuries" in France.[88,89] One British author voiced a preference for the tumbler (a switch that moved up and down just as a light switch seen in most residences in the twenty-first century might), stating that it "deserves its wide popularity in this country, for it is so easy to manipulate; a flick with the finger, sufficing to put it on or off."[90] Critiques of push buttons in other countries took a number of forms. Some, as in the latter case, focused on hand motion; others complained about buttons' expense, and another contingent bemoaned buttons' inability to fit into certain kinds of wall material, such as brick, thereby negating one of the main benefits of buttons: their use as flush or sunken switches.[91] Others still liked the "self-indicating" nature of tumbler or "toggle" switches, which could more clearly show their ON/OFF position in the dark. It is clear that ease of using a switch related not only to technical and practical factors, but also to users' social and cultural familiarity. Push buttons, so common across a range of industries in the United States, connoted magic, pleasure, and gratification as well as simplicity and aesthetic harmony, which appealed to inventors, manufacturers, and advertisers trying to promote electrification.

By 1919, electricians in the United States began transitioning toward the tumbler switch, noting that Americans had previously "stuck" to push buttons despite other countries' lack of interest in the push.[92] In addition to other arguments cited above, one electrician suggested that, "since there is no key to pry loose, [tumbler switches] are mischief- and 'child-proof,'" and that "only a light touch is needed to operate them. If the hands are engaged, the lever switch may be operated at the touch of an elbow."[93] These arguments referenced two interesting facets of digital command: first, they demonstrated that easy access to push buttons by children often posed a problem; and second, they suggested that a "light" or gentle touch could take on varied meanings depending on the user's interpretation. Such points marshaled in favor of one switch over another demonstrate that the practice of pushing buttons was neither natural nor inevitable—stakeholders had to specifically advocate for buttons (and later tumbler switches) as the easiest, most ergonomic, and most desirable choices. Where buttons achieved prominence in some circumstances, they fell out of favor in others.

III Imagining Digital Command

7 What's a Button Good For?

How many of our banks and stores and oil stations have the handy, little, ever-ready button on the job for summoning aid—without the knowledge of the thug—when the unexpected hold-up man appears?[1]

As push-button light switches gave way to toggle or tumbler switches, so too did various constituencies begin to deliberate on the merits and limitations of push buttons in other contexts. In particular, breakdowns in push-button communication in households and offices routinely continued to occur—whether because of the willful disobedience of those receiving a call or user error, prompting modifications to buttons.[2] As in years prior, many expressed indignation about the disparities between manual and digital laborers—those who toiled with their hands versus those who managed the work of others (the pushing and the pushed). Critiques focused primarily on a sense of increasing distance made more prominent by the act of pushing a button as a summons. One concerned party wrote to the *Magazine of Business* (1906) about a printing shop that had "almost been wrecked on such a little matter as a push button."[3] The troubled writer recounted how a manager replaced the business's original owner, who died after thirty years of dedicated service and hands-on interactions with employees. The young manager with his "new ways" contented himself by dealing with the business through a closed office door and a series of buttons at his fingertips—keeping himself out of view from his workers. After a short time in the post, the author revealed, most of the staff quit and a notable set of clients took their work elsewhere. "It was all due to the push buttons," the writer lamented.[4] According to the piece, push-button communication had physically and emotionally separated the manager from personnel and their day-to-day operations, creating an alienating effect. Employees viewed pushing buttons to direct others' movements as inherently disconnected from the former way of doing business that involved a kind of affective "touch" based on human contact.

A fictional account written in the same year described a scenario along similar lines, in which the president of a company, so resolute in his desire to be known only as "the Push Button Man," caused the ruin of a business and himself by virtue of relying on the "whole organ keyboard" of buttons across his desk to keep employees at his beck and call. The story's moral suggested that, "a push button or two on a man's desk are doubtless excusable," but when button pushing came to rule one's way of doing business, everyone suffered.[5]

In part, these tales reacted to blatant nepotism in workplaces that led to easy promotions for some and a sense that newly appointed leaders often let their positions of status go to their heads. Indeed, an editorial (1907) about the call button used to summon railway employees made an elaborate case for the ways that push buttons could inflate a manager's ego, causing him to lose sight of his relationship with and obligation to his employees: "With a finger board nicely arranged with numbered buttons, a certain class of men have become charmed into an exclusive egomania," the author suggested. "The button has, by reason of its magical properties to charm, led some men to forget the courtesies of life and that all knowledge is not confined to any one man, and further, that no matter how wise, there is some one who knows more, and that person may be poor and may be humble." In a dramatic flourish the writer concluded, "Faithful little button! you speed your message, 'come to my office at once,' with no idea of the heart burns and injured pride that may follow your call. Wise is the man who uses your assistance simply to dispatch business, and grows not into vanity as you run your errand."[6] Although the button seemed "innocent," according to this observer, communication in a workplace could quickly snowball into a power struggle at the hands of a puffed-up manager. The author, here, imbued buttons with "charming properties" that seduced their pushers to use and abuse them without consideration of the human beings, whom did not occupy the same social station, at the other end of the call.

Another editorialist came to a similar conclusion, worrying that an employee promoted to a branch manager would give in to his "push button inclinations" and, spoiled by the convenience, would no longer remain an "active fighter for business."[7] Journalist Herman J. Stich (1911) referred to these managers, born out of the efficiency movement, as "push-button gents" and "buzzer-pushing maniacs," known for being "unpopular," "ineffectual," and "impermanent."[8] It is important to remember, however, that this push-button act constituted a privilege offered only to some. Just as employees lamented their role in responding to the call, others remarked at the injustice that only some had access to these buttons. One woman in a newspaper editorial, for example, lamented that, although her husband could push buttons at his desk all day, sending away anyone who bored or displeased him, she asked, "Could I have push-buttons around where nobody could see me push them? Of course I couldn't. Men just have everything their own way in this world, and I wish I'd been born a man, I do."[9] Tension persisted between accessibility and access, as well as between the idealized view of button pushers as docile, feminine creatures uplifted with a mere touch, versus their actual experiences as women in constrained social circumstances.

Whereas buttons could conjure magical effects in other contexts, those buttons pushed for business—that required someone to reply to the summons—triggered continued outrage among the workers put out of view until needed. In the words of one fiction writer (1912):

Well, in these days the gentlemen who are so eager to be very rich have constructed a button—the corporation. It gives them their dearest wish—wealth and power. It removes responsibility away off, beyond their sight. They do not hesitate. They press the button. And then, away off, beyond their sight, so far from them that they can pretend—can make many believe including themselves—that they really didn't know and don't know what the *other* consequences of pressing the button are—away off there, as the button is pressed, people die, people starve, babies are slaughtered, misery blackens countless lives. The prosperous, respectable gentlemen press the button. And not they but the corporation grabs public property—bribes public officials—hires men they never see to do their dirty work, their cruel work, their work of shame and death.[10]

This notion of putting the "dirty work" out of the way by concealing it from view strikingly resonated with a broader project to make push buttons the "faces" of everything good, easy, and simple. The mess, whether it took the form of electrical wires, a servant, or an employee, could remain conveniently at a distance and out of sight. In this regard, push-button communication injured relations because it seemed to put button pushers "out of touch" from those they pushed.

Many communicative problems also stemmed from a lack of feedback, as the person pushing a button often had no way of knowing whether the button conveyed a signal from a distant location.[11] A variety of technologies called "indicators" thus entered the market to create greater transparency and accountability between the button pusher and the "pushed." Indicators often employed vibration as a form of tactile or auditory feedback. For example, one model worked "with a vibrator concealed in the base of the button and makes a clicking noise which is plainly heard by the person pressing the button. If the vibrator is not heard the person will know that the bell does not ring. Much time and annoyance is saved by this little telltale."[12] These systems in some regard allowed the button to "talk back" to the pusher to indicate whether someone heeded the call on the other end, thereby facilitating two-way communication. They also worked to ease the difficulty of communication across distance, where a pusher was not co-located with the effect of her push.

Despite the benefits of increased feedback, some, such as professor of electrical engineering Clarence Edward Clewell (1916), had begun to argue that push-button devices like annunciators "have a common inherent defect" because "they can do no more than to indicate that a party or thing is wanted."[13] He concluded, "If in addition to the annunciator signal it were possible to communicate over the annunciator wires so as to give further information, the value of the system would be correspondingly increased."[14] This perspective became increasingly common, and over time annunciator systems and electric bells with buttons largely gave way to newer technologies like "intercommunicating" systems, akin to present-day intercom systems. It seemed that "the push button is being elbowed from the high office it has long filled in commercial and domestic life," as many began to find that "time is becoming every day more valuable, and there is a distinct saving of it when a commission can be conveyed directly to an attendant, and the preliminary summons dispensed with."[15] Although one-way communication could work as well as—or even better than—two-way communication under some circumstances, in others it posed a liability. Intercommunicating devices in closed-circuit environments thus allowed a user at one end to push a button to make an "automatic" and "almost instantaneous" call, which would allow the person at the other end to reply through a simple receiver.[16] These setups eliminated the need for outside operators and were particularly advantageous when one central room—like a kitchen—could function as a hub between all other rooms.[17]

Although some began rewiring entirely for traditional telephones that could make outgoing calls, electrical supply companies also created and marketed affordable ($1-$2) "push button telephones" as tools for intercommunication in the early twentieth century that could fit into existing push-button bell arrangements (see figure 7.1).[18] These devices "look[ed] like a push button and act[ed] as a push button" but also functioned as a "complete reliable telephone."[19] The user would not even need an electrician to perform the replacement because installing the button phone only required unscrewing the original push button without disturbing existing wiring. Electric supply companies recommended these push-button phones for "interior installation in private houses, hotels, factories and any building where it is necessary to communicate quickly with the various departments this device can be readily installed and commends itself by reason of its small size and economy of installation."[20]

In later years as users adopted such systems, advertisements and informative articles for these telephones emphasized that

Push Button Telephones.

The Push Button Telephone is a wonderful little instrument and represents the latest development in telephone construction, and will give perfect satisfaction when installed properly. It is only 3½ inches in diameter and projects 2 inches from the wall. It looks like a push button and acts as a push button when desired, and at the same time is a complete and reliable telephone. It is an ornament to any room as it can be finished to match woodwork. It can be connected on any bell circuit now in your residence, store, office or factory by simply removing the existing push

Figure 7.1
New inventions offered talk functionality beyond signaling, which adapted push buttons for telephones.
Source: Sears, Roebuck and Co., *Electrical Goods and Supplies*, ca. 1902.
Image courtesy of Collections of the Bakken Museum.

the homeowner could replace the "old-fashioned" button with a "modern" one that could improve communication in one's household without disrupting décor or previous bell-ringing apparatuses.[21] Much like proponents had lauded the electric button for its ability to make housework and communication less laborious, they rallied a similar campaign for intercommunicating devices. Users were promised that they could talk to any part or person of the house by "simply by pushing the proper button" (see figure 7.2).[22]

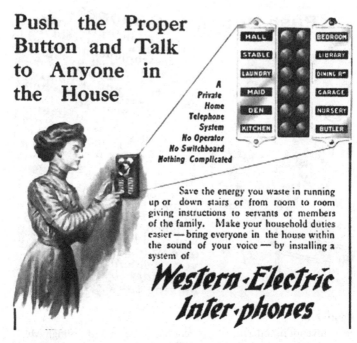

Push the Proper Button and Talk to Anyone in the House

*A Private Home Telephone System
No Operator
No Switchboard
Nothing Complicated*

Save the energy you waste in running up or down stairs or from room to room giving instructions to servants or members of the family. Make your household duties easier — bring everyone in the house within the sound of your voice — by installing a system of

Western-Electric Inter-phones

Figure 7.2
Advertisements for "intercommunicating" telephones emphasized that with a simple push, a lady of the house could communicate with any member of her family without expending effort.
Source: Western Electric Company, "Western-Electric Inter-phones," *Collier's* 44 (1909): cxciv. Image courtesy of Princeton University Library via Google Books.

The Inter-phone and other products like it (such as one popular product called the "Metaphone") capitalized on the concept that a digital commander could accomplish a task with a simple push of the finger, but they took this idea one step further by going beyond signaling through bells and buzzers to the realm of two-way talk. Push button and telephone combinations could potentially make large spaces increasingly intelligible and manageable while making a summons speedier. Indeed, according to the *Electrical Magazine and Engineering Monthly*, "What a convenience to send a message immediately after your signal instead of waiting to deliver it in person. These tiny telephones are cheap, so are within the reach of all."[23] Railroads increasingly made the transition to intercommunicating telephones, too, so that information could be "conveyed directly to an attendant, and the preliminary summons dispensed with."[24]

Intercommunicating campaigns drew on key facets of digital command—reachability (proximity), simplicity, and a gentle touch. Slogans routinely featured reproaches for walking, and instead they advocated for sedentarism in communication as a more efficient strategy. With a Western-Electric "Inter-phone" (1910), for example, the homeowner could save wasted energy from "running up and down stairs or from room to room giving instructions to servants or members of the family" and instead could talk to any part or person of the house "simply by pushing the proper button."[25] Similarly, another manufacturer told potential consumers in later years (1922), "Don't Walk—Push *One* Button *Once* and *Talk*" (italics original).[26] According to the company, "The Stromberg-Carlson Inter-Communicating Telephone System eliminates those needless, wasteful steps. It makes every person, every department in the mill immediately accessible to every other. There is no need to tramp over half the mill to deliver the message."[27] These appeals to efficiency and communication across distance aimed to make pushing buttons a desirable alternative to moving about to conduct one's business.

As with previous push buttons utilized for bell ringing, however, the intercommunicating system relied on users to negotiate how to communicate according to specific social norms; the ability to talk over the wires—two-way communication—did not inherently improve relationships or represent a "better" way of communicating; rather, it merely represented a different way of communicating.[28] As with the transition from bell pulls to bell pushes, usability issues continued to manifest when telephones replaced push buttons in hotels. In one circumstance, a frustrated hotel patron came downstairs to yell at the staff, claiming that he had stood in his room for half an hour pushing a button to ring for a bellboy, only to get no response. The clerk replied that the new system of communication relied on the telephone in the "modern hotel," and that the establishment no longer used the "old antiquated system of so many rings for this and so many rings for that," which prompted the guest to walk away meekly.[29] In another case, a hotel guest recounted:

It was the first hotel I had ever slept in. On the wall of my room I noticed the button and the sign which read, "Push twice for ice water." I was a little scared and lonesome, but I couldn't resist the temptation of seeing how that thing worked. So I pushed the button and stood there for ten minutes, holding the pitcher under it waiting for the water to start running.[30]

Stories such as this one demonstrated that, despite the seeming simplicity of buttons within reach of all, the user required specific social and contextual knowledge to push buttons appropriately. In addition to confusion, two-way talk could also reinforce hierarchical relationships, just as one-way talk could disrupt these relationships when servants or employees refused to respond as requested.

Of note, push-button communication transpired quite differently depending on the forum and the participants involved. Ringing for servants ultimately became less common because of the more widespread adoption of telephones and as a result of changing dynamics related to service and servants in households. However, push-button signaling—building off of early uses of fire and burglar alarms—became more frequent as a tool for safety. Although household communication could be interpreted as unsatisfactory, given the delay between call and response, it represented a boon in environments like schools, prisons, navy ships, or automobiles; button pushing could marry a desire for simplicity with rapid response. The US Navy recognized this utility of pushes as a mechanism for interactions related to emergency. Call bells played a most important part in facilitating communication throughout ships, with circuits for fire warnings, general alarms, warning signals, and routine interior communication; and vessels were littered with push buttons to make signaling effortless.[31] Ships such as the man-of-war "Indiana" were built to showcase the newest and best electrical devices, and electricians often outfitted them with multiple communication technologies ranging from telephones to annunciators.[32] According to a report on electrical installations in the US Navy (1907), every living quarter and office should have these tools, and push buttons would "invariably" serve as the actuating switch for annunciators and bells.[33]

One would find electric buttons throughout the captain's quarters (in his office, above his bed, and even in his bathroom), as well as in staterooms for employees such as junior officers, warrant officers, and the first sergeant. Others who would have access to buttons with buzzers included the navigator, pay-master, executive officer, and electrical gunner. Mess rooms, pantries, and other common areas also provided electrical communication with water-tight pushes. Ships were, in essence, electrical organisms with "numerous telephones, call bells, buzzers, together with a fire-alarm system and the necessary annunciators; the electric thermostats, general signal alarms, electric engine-telegraphs, to indicate the need of an increase or decrease in the number of revolutions per second, electric lamp indicators for various purposes, helm-angle indicators, revolution and direction-indicators, battle and range-order indicators, besides numerous other important devices."[34] Where speaking tubes were once common on these vessels, sailors found that they could not easily understand speech in the chaotic moments when they most needed comprehension. Instead, a push on a button could provide "clear, sharp strokes of a bell without mistake and without occupying too much attention."[35] Indeed, the captain of a ship need not worry about shouting his orders into the wind and could stay securely in "his nest of solid steel" and transmit his orders with a touch, hearkening back to efforts to keep digital commanders at a remove from their employees.[36] Sailors could project warnings with electric horns too, made specially to withstand high-pressure and high-voltage situations.[37] To maintain the many circuits required for button pushing, which connected buttons and bells across rooms and decks, naval electricians were expected to have extensive knowledge of their construction and repair.[38] Among the many adopters of push-button communication tools, the US Navy certainly stood at the fore for its comprehensive use of one-touch signaling.

Around this time in the early twentieth century, other institutions began experimenting with push-button signals for laypersons in case of emergency. These devices combined the communicative purpose of push-button bells with the automaticity of consumer and factory machines. For example, educational institutions began using button-activated "Automatic Electric Clocks" that would sound alarms in every hallway and on every floor; these institutions also began practicing fire drills to prepare for an actual emergency event.[39] Factory buildings likewise embraced fire alarms, with some complex ones that allowed a person to press a button marked "East" (the location in the building of the fire), which would then illuminate the letter "E" on all of the boxes in the building to alert residents.[40] Yet another device married a push button with a thermometer that would automatically give notice should a room reach a danger point from fire. Its manufacturer boasted that not only did the button function as a normal push for routine communication purposes, but it could also save lives.[41] Despite (or perhaps because of) the importance of alarms in these contexts, these buttons occupied a complicated position as "touch" and "don't touch" controls—those who maintained them worried about false alarms, mischief, and misuse. As a result, emergency buttons were often put behind glass barriers. Glass could not only make buttons unavailable for uses deemed unnecessary or harmful, but they could also increase the severity of punishment for those who disturbed them, increasing charges from a misdemeanor to a felony for destruction of property.[42] As with all kinds of buttons, accessibility and reachability often led to a sense of vulnerability that buttons could fall into the wrong hands.

The potential for misuse did not make alarm buttons any less common, however. Numerous button-powered devices that cropped up were meant to serve as signals for police and other emergency officials; in fact, by 1915, inventors filed more than 300 patents for fire alarms alone.[43] Among these, Superintendent Kleinstuber of the Milwaukee police department patented a push-button device that one could affix to a lamppost to contact police either day or night.[44] An article recounting the merits of the mechanism noted that, in addition to everyday uses, buttons could also act as signaling mechanisms for railroad engineers on course for an accident.[45] Fire alarm boxes began to appear routinely on streets and were triggered by a push or pull (see figure 7.3).[46] As they evolved, these boxes were considered "altogether automatic" and could even ring alarms directly in the homes of firefighters.[47]

In domestic contexts, alarms were viewed as indispensable in some homes—primarily wealthy ones, to achieve panicked communication across distance. The homeowner could employ electricity by installing a button in the master bedroom, which, when connected to all the rooms of the house, could instantly flood a room or hallway with light to impede "the enterprising burglar [who] comes-a-burgling."[48] However, these alarms did not exist without problems. Because burglar alarms required syncing up a user's quick reaction with a properly working signal and a response from emergency officials, they were often perceived as unreliable or difficult to manage: "There are as many kinds of burglar alarms as there are burglars—some are good and some are bad, but all of them are troublesome," author A. Frederick Collins (1916) wrote to aspiring amateur engineers in an issue of *Boys' Life* magazine, a periodical for Boy Scouts.[49] In addition to user error, if the button or other parts of the mechanism failed, then the alarm would become "worse than useless."[50] Users marked the threshold for failure of alarms much lower than in the case of other technologies whose functionality did not have such high stakes.

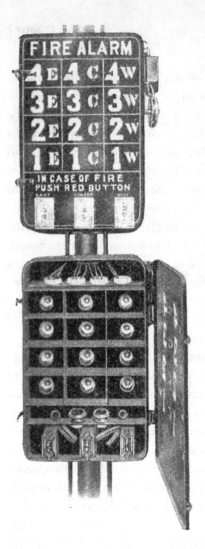

Figure 7.3
Fire alarm boxes featured red buttons, precursors to prevalent "panic" or emergency buttons of later years. These interfaces associated pushes with warning, danger, and instant control.
Source: "Fire Alarm Shows Exact Location of Fire," *Popular Mechanics* 19, no. 6 (1913): 780. Image courtesy of Google Books.

Determining when to push the button could pose a problem for the user; one did not want to "cry wolf" or incur criticism for illegitimate or reactionary behavior. One man learned the hard way when he didn't take a noise seriously in the middle of the night: "'I suppose there are many nervous people, with no greater cause than I have, who would push that button and have the police here in a jiffy,'" he surmised, and decided to ignore the sound and go back to sleep. Much to his chagrin, he discovered in the morning that burglars had indeed invaded the property successfully.[51] The man had worried about inappropriately calling police in the case of a nonemergency; this incident drew attention to the fact that users (and nonusers) interpreted buttons in ways that fit into broader social norms. Yet avoiding the appearance of being "pushy"—forceful despite a lack of force—could come with its own consequences.

Although users often employed buttons to prevent crimes from occurring or emergency situations from spiraling out of control, others co-opted pushes for more nefarious practices. Just as a finger on the button could send a signal that would bring law enforcement officials to one's door, that same finger could help criminals to elude their captors. Journalists often reported on instances when individuals involved in illegal activities pressed a button to signal their compatriots in an act of warning when police were nearby. This strategy most commonly behooved illegal gambling rings, poker games, and pool halls, where large groups would congregate and need to disperse quickly.[52] A "sentinel" stationed outside on the lookout could provide instantaneous warning with a push no different than the one used to ring an electric doorbell, or he could step on a button embedded in the floor that would trigger an emergency red light, taking advantage of the fact that one could conceal buttons.[53] Although it often took police a long time to break up these groups because of this technology, in other cases, the button acted as evidence in officials' pursuit of a criminal. According to a report in the *Boston Daily Globe*, for example, a night patrolman was caught leaving his shift and entering a bank when he accidentally set off an electric bell "connected by wire with a press button fastened on top of the table in the adjoining room," thereby leading to his capture.[54]

The notion of pushing in panic situations took root in the automobile industry, too, with anxiety over how to manage other drivers and pedestrians in a changing transportation landscape that put drivers at a greater remove from their surroundings. Initially, automobile signals resembled bulb horns used on bicycles. These rubber bulbs could cramp one's hand after a time, and they required removing a hand from steering, thus potentially causing accidents. Additionally, after such great overuse, many motorists began to ignore the sound produced by bulb horns altogether.[55] To this end, the electric button alternative could afford a welcome relief. Drivers adopted push-button horns slowly because—as with many electrical appliances—they were often distrusted by the public.[56] These electrical accessories also varied widely in price, with high-end arrangements costing around $35 and less expensive ones in the range of $3 to $5.

By the 1910s, electric horn button products came in a dizzying array of shapes and sizes with a diverse catalog of sounds that inventors designed to make warnings quickly at hand (or at foot).[57] Manufacturers built horn buttons to withstand weather and repeated use, constructing them from rubber or hard black telephone composition.[58] Despite the expense of horn outfits, purchasing only a horn button cost approximately 40 to 50 cents. One could acquire this item from nearly any electrical or automobile supplies catalog and even attach a button to an existing bulb horn.[59] Automobile horns actuated by push button existed as part of a broader landscape of sounds common in the late nineteenth and early twentieth centuries, many of which took on a disciplinary character—a kind of enforcement—meant to convey alerts, warnings, or danger.

Different driving contexts required appropriately matched sounds: advertisers suggested a "powerful warning signal" for city driving and a contrasting "courteous signal" for country and touring trips.[60] Automobile owners could choose the horn that offered the right sound, ranging from the Klaxon , which promised to make enough noise to "awake the dead" (known still today for its infamous "ah-oo-gah" sound), to the SirenO, which provided a "short, quick blast," to the SwarZ Electreed, which was "neither offensive nor musical, just a business-like warning."[61] Although these insistent calls of warning were lauded for their ability to penetrate surrounding noises (e.g., city traffic) in cases of imminent danger, electric horns received highly unfavorable reviews, as did many electric bells used in hotels and homes, from the population at large for the sound pollution they caused.[62] Indeed, the "gentle citizen with tender nerves" would complain about "raucous, ear-splitting" noise.[63] In part, one could trace offensive sounds to drivers as well as enterprising children who would seek out the horn with "unflinching" fingers for amusement, much in the vein of amusing doorbell presses or tricky streetcar pushes.[64] Little boys were often blamed for manipulating automobile horns in their caretakers' absences, with observations that "the horn is 'honked' by every small boy that passes by" and "at nearly all ages boys ... take especial delight in sounding the horns of unoccupied vehicles," which was "extremely annoying."[65] Youthful hands within easy reach of buttons often victimized ears with fingers that ruled the sounds of the day. Most of the time, pedestrians, passersby, and even motorists abhorred horns for the noises they made; although the hand gesture of pushing might have seemed more polite, these sounds were perceived as rude. Still, the cacophony could be handy, for instance, when an offensively loud device with its "piercing shriek," wired to a car door, warded off burglars trying to steal automobile parts.[66]

Alongside signaling to drivers and pedestrians outside of the automobile, push buttons also served numerous functions for facilitating communication between passenger and driver or chauffeur. As was the case with mansions, hotels, or railway cars, only a small segment of the population had access to these devices, and they served to manage relationships (whether successfully or unsuccessfully) between employer and employee; as with push-button managers pushing to command their staff or housewives pushing to order around their servants, buttons functioned as a way of both maintaining and overcoming distance. Electrical products such as the Ever-Ready "Communicator" promised this benefit. Made without metal parts so as not to "soil the gloves and hands of the passengers," the Communicator worked like an annunciator; on pressing a button in the rear of the car, a word with the rider's need would appear on the dashboard (e.g., "left," "right," "slow down," "go faster").[67] The passenger could also ring a bell to gain the driver's attention (see figure 7.4).[68] Electric vehicles appointed with silk curtains, cigar lighters, mirrors, and so on for wealthy passengers also commonly featured a version of the "Communicator" or a speaking tube, as did cabs or broughams to facilitate conversation with the driver.[69] Although these kinds of devices did not

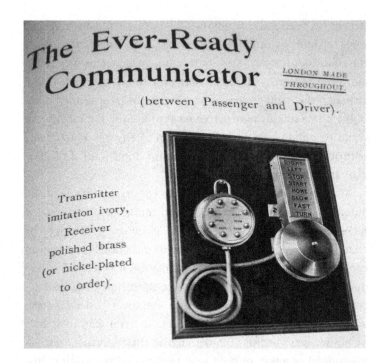

The Ever-Ready
Communicator LONDON MADE THROUGHOUT.
(between Passenger and Driver).

Transmitter
imitation ivory,
Receiver
polished brass
(or nickel-plated
to order).

Figure 7.4
The "Communicator" managed relations between chauffeur and passenger by facilitating communication across distance.
Source: Ever-Ready Corporation, *Eveready Motor Accessories*, 1910. Image courtesy of the Warshaw Collection of Business Americana—Electricity, Archives Center, National Museum of American History, Smithsonian Institution.

achieve widespread uptake, their invention speaks to concerns about communicating within private spaces while disseminating orders in traditionally hierarchical relationships. This act of digital command functioned to enforce such relationships so that the pusher need not lift more than a finger.

An uptick in drivers and sales of automobiles in the twentieth century mandated thinking about modes of signaling while on the road, particularly when drivers and passengers—or drivers and other drivers—were increasingly separated from one another.

Inventors and manufacturers of automobiles argued for the necessity of push-button operation as a luxury feature as well as a safety tool in terms of accessibility. In similar fashion to the arguments made for push-button lights to counteract unwieldy, groping movements in darkness, automobile manufacturers depicted a hand in control and freed from its former exasperations—no "fumbling" of fingers and no "searching" (see figure 7.5).[70] Early electric horns commonly featured a quite simple push-button design; a push button embedded in a steering wheel would trigger an electric bell mounted to the underside of the car's carriage. Pressing the button with the thumb of one's left hand would engage the signal.[71] Although some automobiles would feature a horn in this central location, motorists might choose to set up additional buttons at various places in the car to manipulate the signal.[72] No standardization existed for where horn buttons should be located across all automobile dashboards; this variability sometimes made it more difficult for drivers to acclimate to a new vehicle and understand its controls.[73] Some advocated for "centralized control" as a necessary feature, but wondered, "Will it come? Yes, if the public says so. And then it will be the quest of 'button, button, who's got the

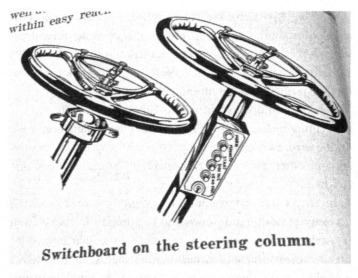

well ...
within easy reac...

Switchboard on the steering column.

Figure 7.5
Buttons embedded in steering wheels or their columns were designed to make control always at hand and to become a seamless part of the driving experience.
Source: S. P. McMinn, "The Car of 1915: Some of the More Important Changes Ushered in with the New Year," *Scientific American* (January 2, 1915). Image courtesy of the Warshaw Collection of Business Americana—Electricity, Archives Center, National Museum of American History, Smithsonian Institution.

button'?"[74] Advocates for placing horns right in the center of the steering wheel looked to the public to generate demand, and they appealed to drivers' safety concerns and need for instant accessibility.

An ad for a Ford accessory horn set the scene for what dangers could befall the driver without an easily touchable and reachable horn:

Thick traffic—a child trying to cross in front of the machine—driver can't reach the horn button in time—and *then!* Too late! Why isn't the Ford horn button placed where it is on other cars—in the center of the steering wheel—where it can be easily, handily reached? This is the device that puts it there![75]

Likewise calling on safety and focusing on reachability, the popular Tuto horn spawned a smaller version called the "Tutoette," which featured thumb control from the steering wheel so that it was "always at instant command and without reaching for it."[76] Advertisements that focused on push-button control for horns emphasized their dependability, quick response, and effortless touch. They told potential purchasers, "One touch of the finger does it," with only slight finger pressure required for easy manipulation of a button that would always be "on."[77] The quality of reachability figured importantly in these configurations because they equated excessive (or any) reaching with danger.

In the case of manual gear shifting, too, formerly the driver was "obliged to take her eyes from the road, release her hand from the steering wheel and grope vainly for a shifting lever, upon the movement of which her very life may depend."[78] However, with a Vulcan Electric Gear Shift, the Cutler-Hammer Manufacturing Co. promised that groping would be relegated to a practice of the past. It didn't matter what kind of hand pushed the button, and economy of that hand's movement—no lost motion—meant that push-button activities were meant to fade into the background of driving experiences.[79] These products and the ads that accompanied them proposed that buttons did not warrant a great deal of the driver's attention, and they did not require frequent repair or adjustment.[80] Manufacturers sold buttons as a seamless part of the driving experience, with the features of digital command as a cornerstone of their appeals. Numerous accessories to automobiles reimagined hand modalities, moving from a model of manual strength to the gentle touch at the tips of the extremities.[81]

Looking at inventors' patents related to push-button automobiles reveals a similar pattern of concern for handiness, reachability, and avoidance of the dangers of groping. Inventors often cited gloves as a barrier to easy button-pushing practices, noting that "it is practically impossible or at least inconvenient to operate the button without removing the glove or mitten from the hand."[82] Designs to remedy this problem made a button larger and more visible so that it "cannot be missed when it is looked for, or felt for, as is more usually the case."[83] Additionally, inventions made buttons increasingly movable so that fingers could strike them from any angle or direction.[84] In the case of automobile horns, a desirable button would respond to "pressure from any direction on the top or sides ... so that the chauffeur can use his elbow, as well as his fingers, to sound the horn."[85] How hands (and other body parts) would push buttons fit importantly into creating new push-button products, as did the question of drivers' attention to the road and the extent to which they needed to see before they pushed. Indeed, "with many of the old type of buttons, the operator always has to take time to look at the button," Ray H. Manson (1919) observed in his patent for a horn operated entirely by touch.[86] Concerns about driver distraction were routinely cited in patent applications, and a widespread belief indicated that touching could take the place of seeing to make the driving experience safer.[87] Beyond automobiles, too, push buttons were viewed as indispensable tools of safety because of their accessibility, such as emergency buttons on factory floors.[88]

As prognosticators looked to the future of automobiles, they imagined digital command with closely grouped controls. Writers for the *Electric Railway Journal* enthused that on learning about recent advances in automobile technology, "We no longer saw an electric car platform cluttered with a controller, brakes, door and step levers, sander rods, gong pedals, circuit-breaker handles and all the other impedimenta that are accepted necessities of the present-day car."[89] Rather, "What we saw in their place was a neat little benchboard on which were buttons or keys with names indicating the several devices, and an attractive young lady seated in a comfortable chair playing on these keys as on an organ!" Describing a cockpit-esque dashboard made for bodily comfort and convenience—with an attractive female operator—this ideal of digital command prioritized both reachability and simplicity.

Likewise, manufacturers further emphasized how buttons would eliminate needless gestures and reduce many hand movements to a single-touch operation. Pullman boasted of its "Little Six" vehicle, "It has been aptly termed the 'Push-the-Button Car.' ... We do not ask the driver to shift gears, start the motor, or do anything but 'Push The Button.'" Referring to the Kodak slogan, the ad concluded, "The gasoline and electric current will do the rest."[90] The company idealized how buttons could reduce a series of manual gestures to one button-pushing finger, delegating the "rest" to the car. To push a button in this context meant not to do anything at all, an appealing notion for individuals intent on eliminating bodily effort from technical experiences. Fetishizing the finger (usually the pointer finger or thumb) served a vital function of reimaging hands that could delegate their labor to others and to machines (see figure 7.6).

This common depiction of an extended finger was repeated across industry promotions, symbolizing the forcelessness of the

**PULLMAN
"LITTLE SIX"**

"PUSH THE BUTTON"

Figure 7.6
An advertisement for Pullman's "Little Six" vehicle emphasized button
pushing as the only skill needed to operate the vehicle.
Source: "Pullman 'Little Six,'" *Motor World* 39 (April 15, 1914): 63. Image
courtesy of University of Michigan Library via Google Books.

digital commander's movements in conjunction with a powerful machine. Advertisers had determined that buttons were
"good for" almost any situation, but in practice users seemed to
accept them more as panic buttons than as conveyors of messages in day-to-day labor relations. Where the urgent nature of
communication via button could justify a forceful demand for
presence—a push for immediate attention—in other circumstances, these pushes represented an injurious imbalance of
effort between pusher and pushed.

8   Anyone Can Push a Button

A learned man has predicted that if the present development of electric labor-savers continues, it will not be very many centuries before we shall become a race of push-buttoners, evolving by slow stages to the point where one large forefinger will be ridiculously over-developed.[1]

Whether pushing buttons for communication or control, an ongoing tension existed among the everyday behaviors of button pushers, public perceptions of them, and the rhetoric surrounding them. It is particularly remarkable that ideals of the "digital commander" remained essentially consistent and unwavering into the twentieth century; conversations about button pushers continued to refer to mere touch, proximity/ remote control, and concealment. The spectacle of "anybody" pushing a button, with an emphasis on the frail, female button pusher, remained a dominant paradigm for thinking about what push buttons could do. A feature in *Cosmopolitan* magazine (1908) on the "world's littlest woman" strikingly portrayed a woman perched on a chair with her finger gesturing toward the button; she posed in nearly an identical position to Mary Newton, the young girl who blew up a mine with a push, and the Otis girl from popular advertisements. The woman could now "do as important mechanical work as the biggest man living" by virtue of pushing a button. The author of the article marveled that the "dwarf" could "direct and control the whole power of Niagara Falls," "release and direct forces that would light streets and homes," and "manage a giant crane able to lift a locomotive or the biggest stone in the pyramids" (see figure 8.1).[2] This framing of the mighty button pusher worked to repair the fragile "other," at once pointing out her otherness while normalizing her physical being to grant her transcendence so she too could harness electricity with a touch.

Another rendering, this one of a hypothetical button pusher of the future, came from artist and illustrator Walter Crane (1907), who turned his talents toward an ergonomic ideal in his "Fancy Portrait of the Man of the Future" (see figure 8.2).[3] In Crane's drawing, the button presser's whole body engaged with electrical technologies. Bells, light bulbs, and speaking tubes protruded from his elongated fingers and connected to his ears. Eight of his toes fluidly pushed buttons, while mysterious tubes of elixir fed into his mouth. The image demonstrated profound interconnectedness between electricity and human; it was almost difficult to tell where the button presser's body ended and the technological apparatuses began. Crane's portrait emphasized this element of the digital command paradigm that involved rethinking where to locate controls (or how to organize spaces so as to maximize reachability) and the ergonomics or "fit" between operators' bodies and the buttons they pushed. Like a theater manager directing the movements of stagehands and actors or William Hammer presenting an electrical spectacle for his friends, Crane portrayed the button pusher of the future as having controls always within reach.

Given the fact that only a limited percentage of the US population would have had access to push buttons in the late nineteenth and early twentieth centuries, this imaginative projection

PRESSING AN ELECTRIC BUTTON
The smallest human creature in the world could direct the
power of Niagara

Figure 8.1
Push buttons elevated even the "frailest" creature and granted her agency, according to rhetoric about electricity and the control mechanism that powered it.

Source: Arthur Brisbane, "The Littlest Woman in the World," *Cosmopolitan* 45 (1908): 324–329. Image courtesy of Google Books.

THE BUTTON-PRESSER—FANCY PORTRAIT OF
THE MAN OF THE FUTURE

Figure 8.2
A creative depiction of a future button presser demonstrates thinking
about ergonomics and the relationship between buttons and proximity.
Source: Walter Crane, *An Artist's Reminiscences*, 1907. Image courtesy
of the Robarts Library, University of Toronto, via the Internet Archive:
http://archive.org/details/artistsreminisce00cranuoft

about what the futuristic button pusher might do, or whom she might be, mattered as much as the practices occurring on a daily basis. Forecasts for push-button life in an age to come were prevalent, and some went as far as to predict that an "ideal future" would have arrived "when life shall consist of sitting in a chair and pressing buttons."[4] Indeed, future prophecies often imagined a time when people could push buttons without getting up, recalling common appeals for sedentarism and ergonomics, as with a fictional story imagining May 1, 2011, when futuristic "John" opened his eyes and, "reaching out his hand he finds an electric button, which he presses."[5] Likewise, in considering how people would come to labor, famous electrician Nikola Tesla predicted that "the work of the future will be mainly 'touching electric buttons' to set automatic machinery in motion."[6] These proclamations, some wildly romantic and others quite bleak, offer insight into how users reevaluated their hands in a world increasingly powered by buttons.[7] They transitioned from thinking about individual users (e.g., managers, housewives, or amateur photographers) as button pushers and instead envisioned a race or nation of people who pushed buttons.

These prophecies circulated regularly, in part, because of heady and pervasive emphasis on grooming ideal button pushers—especially as workers operating machines in factories—in the midst of the scientific management movement's unbridled enthusiasm for "efficiency." Efforts to coordinate cooperation between hands and machines abounded, and those who studied button interactions in scientific and medical industries tried to make sense of how users could harness pushes to streamline labor, make reaction time quicker, and harness touch for greater economic impact. Questions circulated about who would make the best button-pushing worker; as discussed previously, children often served as effective test cases for buttons to demonstrate their accessibility to the masses. In fact, some deliberated over whether children could become the primary workers in factories if they could carry out work by button inexpensively and with little training or strength: "When we push the button ... and set in motion not only a machine but an entire factory, we are prone to think that all that is necessary is to secure a child to do this," argued J. H. Morgan in a case against child labor to the state of New York in 1903.[8] Concerned citizens aimed to protect children from factory work as part of a broader movement away from child labor in the face of redefinitions of childhood, in part by arguing that the simplicity of pushing a button did not make youth employment morally responsible.[9] In arguments against easy, effortless labor, a puritan work ethic dominated discussions, with assumptions that work should be difficult and performed by skilled workers rather than by the most undeveloped and untrained members of society.

As in other circumstances, push-button evangelists' lofty ideals of button pushers were out of sync with actual practices. Children often proved incompetent at using buttons, demonstrating that simply being able to push a button differed significantly from using a button safely and appropriately in all contexts. In one much-discussed case on May 13, 1904, young Dorothy Fisher caught and crushed her leg in her apartment building's elevator. Fisher, who was five or six years old at the time of the incident, won a lawsuit against the owner of the building when the lawyer argued that the operator should have taken measures to protect children from the automatic elevator, an "enticing plaything."[10] The lawyer cited a previous Minnesota decision (Keffe v. Milwaukee & St. Paul, Ry. Co.), which stated, "When it sets before young children a temptation which it has reason to believe will lead them into danger it must use ordinary care to protect them from harm."[11] In these cases, the summary concluded, children "of tender years could fail to know and appreciate the risk," and therefore required protecting. As a young girl, Fisher no doubt received additional sympathy, and her portrayal as a victim suggested that her naturalized "temptation" to push the button existed beyond reproach. The system had to grapple with cases such as Dorothy Fisher's, in part because a chasm existed between representations of young girls as innocent, empowered button-pushers and the realities of their behavior. Girls were expected to perform according to social expectations of ladylike behavior—and to naturally understand how to push buttons without incident—and thus their antics received little formal discussion. By holding children up as totems for the effortlessness of button pushing, the electrical industry had created an ideal user who represented accessibility but threatened in terms of access. Although it is unsurprising that advertisers relied on themes of effortlessness, magic, ease of use, and "touchability," all tropes that predominated in promotions of electrification more broadly, it is telling that a "control revolution" was so routinely characterized by bodies performing control in ways that deviated profoundly from such representations.[12]

As a result, the more that bodies performed "out of control," the more efforts ensued to create protocols for managing button pushers—especially when it came to the world of work. Scientists, physiologists, and management enthusiasts viewed efficient button pushing as an integral part of a well-functioning, physically optimized labor environment. Such efforts fit in with goals at the turn of the twentieth century, according to one contemporary scholar (2011), to create "a coherently articulated science of touch" and to use touch to "quantify and monitor workplace fatigue, ensure the proper cultivation of the senses in pedagogical programs, and enable a host of other psychological diagnostics."[13] Push buttons functioned in many ways as tools of rationalization and discipline when users employed them to control workers' hands—so that they would "follow the rhythm of the mechanical system" and direct their physical forces toward desired ends.[14] These endeavors played a part in schemes across nearly every sector of society to establish widespread bureaucratic measures and "scientific" practices (in medicine, schools, prisons, etc.) and to streamline and control numerous aspects of daily life.[15] Forms of discipline—encouraging hands to work in certain ways, for example, worked directly on bodies, coordinated their efforts, and managed their desires.[16] In this regard, buttons operated as mechanisms to advance institutional and state aims.

Experiments took a variety of shapes, but many investigations focused on button pushing in relation to reaction time, studying how quickly a human finger could touch a button under varying conditions. From a scientific perspective, reaction time merited attention because it offered insight into the "bond of relation between mind and matter" because "all material actions require[d] time."[17] According to University of Illinois Urbana–Champaign professor of psychology William Otterbein Krohn (1895), one could study reaction time quite simply; he proposed an experiment in which a subject would push an electric button as soon as the experimenter touched the subject's forehead with a pencil. Krohn surmised that, although the experiment required little in the way of materials or special conditions, reaction time was a "very complex affair" that entailed the sense organ's impression when the forehead was touched, impulses triggered from nerves to the brain, transmission of the motor impulse to the finger, the contraction of the finger muscle, and the finger actually reacting and moving.[18] After cataloguing this physical process, Walter Moore Coleman (1905) concluded that it took a person an average of an eighth of a second to press an electric button after seeing a signal.[19] Another study years later by Michael Vincent O'Shea and John Harvey Kellogg (1915) determined that it took about one tenth of a second to see a card placed in front of one's eye and to touch an electric button indicating that one had seen it.[20] Buttons became the de rigueur method for studying reaction time, with interests in identifying the optimized hand according to its minimal finger movements.

These studies, in addition to satisfying the curiosities of their testers, often functioned to fuel researchers' particular agendas. In the case of O'Shea and Kellogg (1915), the authors advocated that, "The chief thing for any person is to study the question of making his body a smooth-running and effective working machine, so that it will always be ready for any task or enterprise. Only in this way may the most be got out of life."[21] These individuals believed that honing reaction time—and making one's interactions with communication and control mechanisms effortless—would aid the human body by enabling it to achieve its full potential.[22] Well-known researchers such as Frank and Lillian Gilbreth and Frederick Winslow Taylor, the latter an advocate for a brand of scientific management known as "Taylorism," desired this optimization to make laboring bodies, especially in factories, more efficient. In his classic book, *The Principles of Scientific Management* (1911), Taylor noted that university physiological departments routinely carried out experiments with electric buttons to "determine the 'personal coefficient' of the man tested."[23] He enthused that some

individuals were "born with unusually quick powers of perception accompanied by quick responsive action."[24] To know each
worker's "personal coefficient" and assign him to his most effective post could help a manager to maximize production from
his workers' bodies and therefore improve his business. Impedi-
ments to reaction time, such as fatigue and attention span, were
cataloged to understand the full range of human responses
possible and to subsequently mitigate them.[25] Conversely, this
information also worked to reduce the amount of human labor
performed, as in cases made to shorten the length of workdays.[26]
Electric buttons figured importantly into these tests as common
tools for measuring the effectiveness of manual machine interactions (see figure 8.3).

To control the labor force, these efforts not only worked on
bodies as a whole, but also broke them down into parts and
focused on the hand as a site of intervention. Teaching the
hand to operate with only a digit could make bodily movements
streamlined for rapidity of action and increased production.

Yet the pressures of factory work, of behaving efficiently with
buttons, often took their toll. Workers liked to push the boundaries of what buttons could do, and managers often responded
in kind by trying to create a button resistant to tampering. Button pushers received admonishments for jabbing buttons with
materials other than their fingers or for creating ways to keep
push buttons held down while they performed work with their
feet.[27] Some workers even lost fingers or hands in their "misuse" of push buttons out of desperation because they "needed
the money" and wanted machines to work faster. Factory managers also complained about "thoughtless" machine operators
who allowed their fingers to remain on the button longer than
necessary.

HIS LEFT HAND WILL BE TOUCHED; HIS RIGHT WILL PRESS THE BUTTON

Figure 8.3
Measuring humans' abilities to react to electrical forces, physiologists, psychologists, and scientific management experts typically used electric buttons as the most common mechanism for this process. As was often typical at this time period, children served as "ideal" users and test subjects.
Source: Frances Gulick Jewett. *Control of Body and Mind* (Boston: Ginn & Company, 1908). Image courtesy of Harvard University Library of Medicine via Google Books.

In other instances, adults also played and caused mischief and were known to "over-push" buttons, such as mechanics who liked to take dumbwaiters (elevators) for "joyrides."[28] Meanwhile, inventors disparaged the "rainy-day customer" of elevators who would "pass the time by poking at the push buttons with an umbrella."[29] Even in prisons, officials debated whether prisoners should have access to push buttons to flush toilets, noting that some station houses in New York had witnessed prisoners "interfering" with the buttons.[30] In this case, the reporter recommended ensuring that buttons were installed properly—flush into walls—so that button pushers could not easily disturb them. In public environments, especially, issues of control frequently cropped up to make buttons as resistant to tampering as possible. For buttons used to trigger alarms and in emergency situations, inventors suggested monetary fines for indiscriminate pushing. Building administrators also routinely put them behind a pane of glass to avoid "meddlesome people."[31] Locked buttons appealed to users in places like office buildings and asylums, where access mattered most and mischief and destruction of buttons often occurred.[32] The Anchor Lock Push Button Switch, for example, provided an attachment so that only a user with a key could activate the button, thereby thwarting those individuals who, "through ignorance or a wanton disregard for consequences," might take advantage of easy access.[33] Fooling around with push buttons necessitated increasingly complex button arrangements to outwit the "fools" who used them.

Indeed, around this time and within this context, the term "fool-proof" also began circulating as a philosophy of designing products for both workers and the masses. Whereas previous slogans for products had focused on reducing effort or previous skill, to make something foolproof went a step further because it viewed the user as inherently incompetent. This perspective led to the design of machines and products that might augment human limitations and protect supposedly stupid, lazy, or untrained users from themselves. According to advertisements and trade magazine articles, the "fool" could take various shapes, such as a housewife without competencies to use a machine or a "careless" factory worker previously prone to mistakes and injury.[34] Some worried that the term "foolproof" leveled a "'slam' at the intelligent" and instead called for campaigns and inventions targeted at the brainy, capable consumer.[35] Still, a view of users as fools predominated, with buttons occupying center stage as panaceas to simplify consumption. Some went as far as to champion "fools" as finally "free" because merely pushing a button could unburden them from having to understand the machines they used:

We are all becoming more and more addicted to the pushbutton habit. ... In industrial work the man with the push need not, and often is not expected to know what happens, for instance in a motor drive, between the button and the motor. His attention is devoted to the performing of a certain operation, with the motor as the propelling force. After he has indicated his desire for motor power by pushing the button he can turn his attention without distraction to the work in hand and leave that relatively unknown territory between the end of his finger and the delivered mechanical motion to the electric man.[36]

Here, the "unknown territory" beyond the fingertip need not concern the worker, an argument common in debates over automation and delegating the effort or thinking to machines. Without the need to think or train one's hand, perhaps "physical strength is no longer an element, a woman can pull a lever or press a button as well as a man, and the aristocracy of sex disappears. A black, brown, or yellow man is as good an adjunct to a machine as a white man, and the aristocracy of color vanishes."[37] Although these ideas about button pushing as an "equalizer" could have had democratic implications, opening up a realm of access to those previously disenfranchised, they instead threatened ingrained definitions of humanness. Which characteristics distinguished humans from machines—or humans from other humans, for that matter—without the hallmark measurements of strength, skill, or difference?

Buttons served as the ultimate symbol of a society wracked with uncertainties about their employability and individuality, where "mere touch" could also imply that people represented mere "hands" or adjuncts to the machine and nothing more. Indeed, as one author worried, "All there is left for [a skilled artisan] to do is to pull a lever or press a button; and by this process the separate craft lines are obliterated and all workmen are leveled down to the plane of operatives."[38] Increasing mechanization led many to note that "muscular exertion is less and less a factor in the accomplishment of human aims," and worries abounded that manual labor would become a thing of the past. They also railed against a sense of button pushing as nonwork. One writer complained, "Open plumbing and push buttons have destroyed the manifestation of those Spartan traits that make for self-confidence and independence."[39] Button pushing represented labor-gone-"soft," and onlookers positioned it in opposition to those individuals, almost always of lower position and class, who "made" themselves by clawing their way up from nothing. Author and soon-to-be Socialist Party candidate Allan Louis Benson made a similar argument in his essay, "If Not Socialism, What?" (1914) Benson described a future world in which one wealthy tycoon could own a "wonderfully automatic" machine that he would control from Wall Street and that would be "perfectly operated by pushing a button."[40] The owner would have an embarrassment of riches, Benson surmised, whereas "nobody else would have anything."[41]

Such anxieties, which stemmed from sometimes overzealous or oppressive efforts toward efficiency, led to ongoing complaints about button pushers—digital commanders—as unskilled nonworkers made spectators by automatic machines. The too helpful machine threatened to make workers obsolete in the name of efficiency. Likewise, decades after the practice of ringing electric bells began, manual laborers who did not have access to push buttons continued to harbor resentment about those who "labored" with their fingertips. A joke in *Life* magazine (1922) satirized these feelings by beginning with the following question: "How do millionaires get their daily exercise?" The punch line proposed that the well to do might use their hands to push buttons to burn a few calories:

Have large flat-top desk in private office fitted with row of push-buttons within convenient reach of your chair. Inhaling slowly, instruct your secretary to carry chair around to opposite side of desk. Lean forward flat across ink-well and push buttons, calling in turn Production Manager (exhaling), Sales Manager (inhaling), Efficiency Engineer (exhaling), Building Engineer (inhaling), and Special Cop in the main hall (exhaling). Note: Care should be taken to inhale only after exhaling, otherwise half the benefit of this exercise will be lost.[42]

It would take more than a few breathing exercises and finger punches to work up a real sweat, *Life* magazine surmised, mocking the supposed "efficiency" of hierarchical, bureaucratic institutions and the hand practices they made popular. Whereas electrical manufacturers and electricians promoted digital command, in workplace contexts, the act of pushing buttons by managers intensified the physical, economic, and social distances between workers, with unequal hand gestures representing yet another layer of stratification.

Similarly, in the case of a fictional story titled "The Push-the-Button Man" (1924), the protagonist—enamored with the prospect of sitting behind a desk and pushing buttons all day—ultimately had to excise buttons from his life to reclaim it. The story dramatically unfolded as:

He sat down at his desk and his eyes rested upon the block of call buttons which had so nearly proved his undoing. There were twelve buttons; white, black, and red. Carey took out his knife and cut the cord that connected them with the other desks. The block fell to the floor;

Carey picked it up, took a sheet of paper from a pigeon hole and wrote: "I am sending you the buttons from my desk, as I shall not want them any more. Keep them as a souvenir of the past. Will see you tonight and explain."[43]

Here, the main character, Carey, violently did away with buttons by "cutting the cord," both literally and figuratively, to achieve salvation. He then faced his future starting out at the bottom of the company and working his way up, noting, "It was going to be worth while; out there, where men made themselves; where they held their destinies in their own hands; where merit, alone, won; and—where no one ever pushed buttons."[44] At the end of the story, Carey awakened to the error of his ways and embraced the ethos of the "self-made man." This moralistic tale and others like it demonized push buttons as a way of indicting a growing rank of digital commanders whose management style, from small hand gestures to working at a distance from other employees, often caused alienation. Although anyone could push a button, not anyone could have access to that button, and therein lied the rub. The privilege associated with button pushing meant that some seemed to "push" their way to the top without effort.

9 Push for Your Pleasure

Although button pushers garnered negative attention in working environments for their entitled, lazy, and forceful ways, electrical enthusiasts continued to invest in a vision of the user as anyone, which replicated over and over again in the pages of newspapers, magazines, electrical catalogs, and manufacturers' advertisements. In similar fashion, the electrical industry sold electricity and automatic machines as the perfectly responsive servants whom would solve every button pusher's woe, much in contrast to the unreliable or outright angry human and electrical attendant. A *Washington Post* article (1907) asked its readers to fast forward 100 years to 2007 AD. At that time, people would live in electrical houses that automatically cleaned your shoes. An automatic vacuum would remove every particle of dust from your clothes as you entered, and an automatic oven would cook roast beef and vegetables, alerting the homeowner when the food was prepared. Push buttons would activate searchlights that could follow any visitor on the property, determining automatically whether to grant entry. In this arrangement, the author reported, "the domestic problem is going to be almost entirely eliminated in the house of the future."[1] Indeed, "Electricity, properly applied, has already proved a most capable servant."

As the previous chapters have discussed, the original function of push buttons primarily involved calling servants or other
employees to action. Buttons operated according to the logic of
"beck and call," where the pusher's finger beckoned a person or
thing desired so it might appear. However, as the middle class
could hire fewer servants for domestic help, and as scientific
engineering and domestic management movements increasingly prioritized "efficiency," society had begun to think differently about servants and service. Such a shift in labor and hand
practices in the United States transpired for a number of reasons,
including changes in technology, evolving roles for domestic
workers, and an increasing focus on "self-service" as a do-it-yourself model of consumption.[2] People started to employ machines
as "performers of work of low social status," in the words of
one recent scholar (1995), who has noted that a dominant way
of thinking took root that "some being or other must serve to
make it easier to live and work."[3] Some Americans demonstrated
pride in replacing human servants with electricity as a symbol
of modernity and progress. Indeed, an article on "Calling the
Servants" in the 1920s noted, "Now we push a button, but a century and a half ago handbells and whistles were used to call a servant, and in an old comedy of the reign of Charles II of England
the company was called to dinner by the cook knocking on the
dresser with a rolling pin."[4] Creating a bridge from the servant
paradigm to the electrical servant paradigm involved rethinking
how people would use their hands to accomplish the tasks of
everyday life without increasing the burdens they felt. Not only
did approachable buttons offer a captivating face of electricity
that experts believed could seduce uncertain nonusers, but button pushing could also provide a kind of seamless transition to
labor carried out by electrical technologies rather than people—
just "call" for anything you desired and a push could make it so.

Unlike human help, electrical servants never "ask[ed] for a day off."[5] This philosophy dictated that the machine was "a kind of non-human slave, tireless and nerveless"; in interactions with that machine, "Man's part is perfunctory: to pull a lever, to push a button, or to turn a crank; the more automatic the machine becomes, the less is there need of man's assistance."[6]

Tropes of "magical" electricity relied on the notion of having electricity always on demand.[7] Push buttons could only fulfill their promise to deliver at one's beck and call if the thing requested appeared within an instant. In fact, electric utility companies developed a "readiness-to-serve" minimum charge in the early twentieth century for electricity users so that electricity would be "holding itself in constant readiness to render service" and "ever ready to meet the consumers' caprice."[8] However, these companies faced some challenges in selling consumers on the idea of paying for a service they may or may not use. As a result, electrical utility companies hoped that thinking about electrical readiness or on-demand electricity through the analogy of servitude would make a more compelling case for paying a readiness fee. In a perfect world, pushing a button should deliver effects instantly at will, just as a servant should arrive speedily upon being called, or the concept of "push for this" or "push for that" would fall short of its promised benefits.

In 1916, for example, the Society for Electrical Development, Inc., chose a poster for "America's Electric Week" from among 781 entries that whimsically celebrated the benefits of button-powered electricity (see figure 9.1).[9] An electrician, writing a positive endorsement of the poster enthused, "Gone is the ancient lamp. Now it is the gentle touch of a button and forthwith comes the Genie, Electricity."[10] The campaign and its admirers indicated that users could conjure up electric miracles as long as a button stood at hand. The push heralded electricity

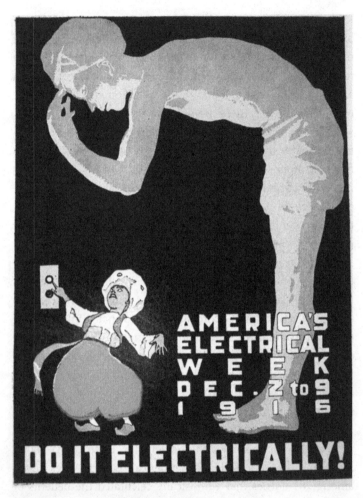

Figure 9.1
Poster depicting push-button interfaces as magical and instantaneous.
Source: *Electrical Review* 69, no. 8 (1916): 321–322. Image courtesy of
the Warshaw Collection of Business Americana—Electricity, Archives
Center, National Museum of American History, Smithsonian Institution.

seemingly from the heavens, making electrical circuits, wires, plugs, and other mechanisms invisible.

Imagery of genies springing from buttons abounded. In fact, the National Electric Light Association's (NELA) electricity "mascot" named "Kilo-Watt" appeared on a number of promotional pamphlets as the "Genie of the Button" to increase the public's adoption of electricity (see figure 9.2). These popular depictions of electricity as a genie ready to grant one's wish made the connection between servitude and electricity clear.

In another instance, Loring Pratt (1917) wrote a romantic ode to push-button electricity that celebrated the "invisible Genie":

Out of the Chasms of Chaos
Out of the Great Beyond
Down from the heights of Olympus,
Forming an earthly bond;
Never yet seen by mortals,
Ever surrounding all –
Patiently, potently waiting,
Ready for every call.
Thus does the mighty Genie,
Coming from realms beyond,
Silent await our bidding –
Silent and swift respond.
Chained to a little button,
Ever in reach of your hand
Stands this invisible Genie,
Always at your command.
Only a little button
Holding a Giant's power,
Turning the wheels of business,
Lighting the darkest hour.
Only a little button
Ruling a Giant's might,
Only to press that button
For Heat and Power and Light.[11]

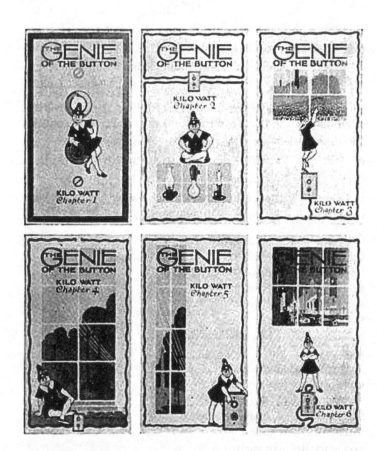

Figure 9.2
"Good will" brochures designed by the National Electric Light Association sought to make a compelling connection between push buttons and genies as the magical purveyors of electricity.
Source: *Electrical World* 77 (1921): 834. Image courtesy of Google Books.

Pratt's poem emphasized a number of facets of idealized digital command—the little button stood "ever in reach of your hand" while making invisible or intangible electrical energy entirely within one's grasp and perpetually ready at a touch. Much like the servant who preceded it, electricity could remain out of sight until needed. As discussed in chapter 3, this strategy of putting "the help" out of sight reflected "The history of domestic service in the United States," where "the social role 'servant' [has] frequently carrie[d] with it the unspoken adjective *invisible*" (italics original).[12] Covering up electricity behind push buttons continued a tradition of putting servants in a position of limbo in terms of their visibility or, in the words of one servant, "being there and not being there."[13] Electricians at once promised the electrical servant as an ever-present helper and yet thoroughly unobtrusive.

Interestingly, this master-servant dynamic by button—which prized strategic visibility and invisibility—was often described as a particularly American phenomenon. In other countries, such as India, an electrical goods dealer from the United States (1922) noted that, "Servants are numerous and are supposed to be within call when wanted. It was noticed that even in some of the large hotels bell systems were not provided, the theory being that everyone travels with a personal servant or 'bearer,' who generally sleeps in the hall near one's door, so that bells are unnecessary." Likewise, Jacob Warshaw's study of *The New Latin America* (1922) referred to buttons as the epitome of American efficiency and therefore quintessentially American, for "[Americans] push a button: [A Latin American] orders a servant. We turn on the heat: he tells a servant to make a fire or to stir it up."[14] The desire for invisible servitude, then, fit within a US cultural paradigm that involved keeping servants—and the electricity that replaced them—out of vocal range and sight.[15] Digital command by button appealed more to those determined to keep workers at a physical remove.

In this context, then, the image of the invisible servant heralded by button continued to circulate. An advertisement for the 1920 New York Electrical Exposition sought to illustrate how humans could take control of electricity to do their bidding through digital command. The poster, designed by Fred G. Cooper, "shows Electricity as an untamed element, in the process of subjugation. A vivid flash of lightning dominates one side of the poster—two hands typifying human skill, have gripped this flash, tamed it, harnessed it to a push button and put it to work."[16] Here, it is particularly noteworthy that—as with Pratt's poem—hands functioned as the implements for "subjugation," reducing the wildness of electricity to a push button so that it might become a slave who performed on command.

Even when the metaphor of magic or invisibility did not apply directly, the electrical industry continued to promote a view of buttons as servants. The *Edison Monthly* (1922) cheerfully proclaimed that:

Just as it helps the wife of the bachelor's more fortunate friend and the bachelor's sister in solitude, the bachelor-maid, in the accomplishment of their household tasks, so the push button helps the bachelor in his singular housekeeping. As pointed out elsewhere in this issue, the push button has become a veritable valet in the home of the bachelor— lighting his den, preparing his shave, cooking his meals, caring for his health, and in general making bachelorhood so comfortable an institution that it is to be feared nothing can lure him from it.[17]

Interestingly, the wording of this passage referred to the button as a kind of anthropomorphic servant—it was the button (not electricity or a human servant) that took care of lighting, cooking, and caring without complaint.

The metaphor of electricity as a servant extended beyond push buttons, and it commonly referred to a desire to "tame" machines and their properties with humans' powerful hands to establish a master/servant or dominant/submissive relationship.[18] Indeed, "At your beck and call, day and night, always and forever, stand the great public utilities—they are your obedient servants," one article in *Electrical World* enthused.[19] Northwestern University president Dr. Lynn Harold Hough (1920) agreed, writing in a piece titled "The Age of Machinery" that, "It is a good thing to use a machine as a slave."[20] Advertisements such as one for a Herrick refrigerator proclaimed, "The Unseen Servant behind the Perfect Meal Is the Perfect Refrigerator," while headlines in newspapers foretold of the bright electrical future on the horizon: "Electric Slave to Serve Homes."[21] Although electricity functioned like a servant, it was specifically "with a button or plug we have made it our servant." In this view, buttons acted as disciplining or harnessing forces that created the possibility of subservience. In other words, buttons *made* electricity a servant.

Such fixation on the tireless electrical slave stemmed in part from a concerted effort on the part of electrical goods manufacturers to sell the practice of pushing buttons as not just a curiosity but a way of life. Given that touch often connoted pleasure—whether sexual, intimate, healing, or otherwise, electrical companies drew on this connection to associate the push of a button as a feel-good practice, as with slogans such as, "It's a good thing: Push It!" or the Edison Company's recommendation for the benefits of push-button lighting: "If you are depressed, touch a button and have a few volts of cheerfulness."[22] Push buttons should afford both creature comforts and physical relaxation; they should put one's desires "within reach" at all times. In appeals made to consumers, buttons symbolized a growing emphasis on consumption as pleasure and everyday technologies as a vehicle for such gratification. The push-button culture quickly grew entangled with consumer culture. Good consumers should push buttons to get what they wanted, so these messages implored.

Although the public may have reacted with uncertainty about doing away with the human intermediaries who had previously carried out tasks, push-button consumption had gained traction through the introduction of vending machines, elevators, cameras, and light switches in the late nineteenth and early twentieth centuries. Given that many of these technologies were but prototypes—not widely used by the masses at this time period—they functioned as "dream machines" of their day—visions of what life could look like in the future when an individual could manage her business dealings, relationships, information needs, and emotions solely by pushing.[23] The ideal button pusher desired and sought out convenience, and she wanted what she wanted when she wanted it. Just as anyone could (potentially) push a button, as discussed in chapter 8, it became increasingly common not to think about a particular desire being satisfied—such as pushing a button to dispense candy or take a photograph—but rather to think about *any* desire being satisfied by buttons all at once.

In 1905, Professor Charles Henry Rieber of the University of California hoped to develop a machine of this kind that could provide its user with anything she desired, including answers to her every question. Author W. B. Nesbit—inspired by Rieber's efforts—pondered the possibilities of this push-button logic in an intriguing and prescient poem titled "Push the Button":

If you have a lot of questions and are
worrying for answers,
Push the button.

If you want to know the ages of the youthful ballet dancers,
Push the button.
If you want to know the reason for a vexing lot of things,
The worries of the commoners and discontent of kings,
And why you have to coax the youthful
wonder ere she sings,
Push the button.
Would you get some information as to
stocks or bonds or grain?
Push the button.
Do you want to make a million? Do you
want to catch a train?
Push the button.
Is it something as to fashion—as to bonnet, gloves or dress?
Does the Christmas problem worry you
and fill you with distress?
Do you wonder if the damsel will reject
you or say "Yes?"
Push the button.
For the price of lamb or lobster, beef or
veal or pork or mutton,
Push the button.
For the work of all the authors, from old
Chaucer down to Hutton,
Push the button.
Now the information bureau is a thing of
cogs and wheels.
At the shifting of a lever all its knowledge
it reveals.
If you'd like to know the outcome of your
doings and your deals,
Push the button.
There's a button for your likings, for your
longings and your joys—
Push the button.
And you needn't go to school so long as
you've got the strength to shove—

Push the button.
If you want to ask a question, and the
dial don't reply.
If the thing is out of gearing and its
works have gone awry.
Do not go to the inventor and in anger
ask him why—
Push the button.[24]

In a striking piece of writing long before computers, smart-phones, or "apps," Nesbit described a machine that could satisfy any want—one's "likings," "longings," and "joys," the status of one's relationship, as well as information on any topic—with the only requirement being that "you've got the strength to shove."[25] The scenario captured an already pervasive fixation with buttons, which glamorized a world where fingers operated effortlessly. In Nesbit's example, the push-button experience was totalizing and all encompassing, with one hand practice made to accommodate every need and whim.

This vision of a complete push-button existence wasn't always viewed positively, as in the case of E. M. Forster's dys-topian short story, "The Machine Stops" (1909), where buttons could procure anything the protagonist desired, but the act of pushing would ultimately lead to the button pusher's severe and profound isolation:

There were buttons and switches everywhere—buttons to call for food for music, for clothing. There was the hot-bath button, by pressure of which a basin of (imitation) marble rose out of the floor, filled to the brim with a warm deodorized liquid. There was the cold-bath button. There was the button that produced literature. And there were of course the buttons by which she communicated with her friends. The room, though it contained nothing, was in touch with all that she cared for in the world.[26]

Rather than depicting an empowered consumer gratified by her consumption, Forster feared that this automatic living—push-button living—would alienate and disconnect a button pusher whose only ties to the rest of the world were buttons. Such a critique viewed long-distance, technologically enabled communication as a barrier to self-fulfillment rather than the hallmark of modern and efficient existence. Here, as with other concerns about distance and alienation, Forster proposed that being "in touch" via a push-button universe was not the same as the closeness of "human touch."

Yet another critic focused on the mindless pleasure and "mania for simplification" brought about by pushing buttons, where buttons would produce a dangerously "effortless state":

It is no longer necessary to speak to be served. You step into a hotel, press the button, and a succulent luncheon appears suddenly before your delighted vision. Ten seconds later you feel chilly; you press another button, and presto! Your fireplace is lighted up as if by magic. Electric buttons have become the masters of the world, overcoming distance, doing away with the necessity for forethought, and, for that matter, for thought at all. Everything is changed.[27]

As with Forster's example, once again the role of distance featured in this account; proximity and distance could take on different perceptions and meaning. Sometimes closeness felt too close, and in other cases distance felt too great. In this instance, the author romanticized effort as a form of more authentic engagement with the world than the effortlessness that made overcoming distance possible.

These utopian and dystopian renderings of push-button life began to take shape for electricians, manufacturers of electrical goods, and advertisers, whom often brainstormed ways to most effectively stimulate desire in consumers and create a need for their products. Joseph French Johnson, dean of the NYU School of Commerce and president of the Alexander Hamilton Institute, argued to advertisers (1920), "So if we want to make men more civilized we must make them want more things."[28] Describing how advertising could serve as an "economic force" to encourage these wants, he illustrated a fictional scenario where one could link buttons to desire and gratification:

Suppose an architect could put into a man's house a room on the walls of which were a myriad of electric buttons, the pushing of which would bring the gratification of any conceivable desire—here a button that brought out a beefsteak dinner, there a button that gives him a bottle of red wine, over there a button that responded with cigars, and here buttons that responded with music of any kind desired, or with beautiful pictures, or with books and magazines that exactly suited his fancy. In that magic chamber there would be, of course, some buttons, perhaps many, which good people would call wicked buttons; for example, in the opinion of some people, the wine button or the whisky button. Now if any architect could construct that kind of a room you know perfectly well that his services would be in constant demand even though he lived forever.[29]

Johnson's imaginative rendering of a world ruled by buttons demonstrated how advertisers thought about pleasure as a manifestation of goods available on demand. In this universe, a push button existed for every want, whether good or "wicked," ready for the button pusher and always at hand. Rather than summoning one servant to bring each of these items, every item would have its own button ready for dispensing.

Storeowners often tried to create and enact this kind of push-button experience (in a more limited way) for consumers—or potential consumers—to entice them and offer a form of unthreatening exposure to serving oneself by button. To encourage consumers to consider purchasing a washing machine, for example, the Tremaine Electric Company in Brockton, Massachusetts, recognized that, "The majority of grown-ups enjoy 'seeing the wheels go 'round' and very few of us are averse to pushing the button that starts them going."[30] Thus, they "took this little characteristic of human nature into consideration" and designed a push-button window display for pedestrian use.[31] Likewise, the Electric Construction and Fixture Company of Savannah, Georgia, put out a window sign for passersby that read, "If Hot, Push the Button" to activate a fan as they walked by the storefront.[32] A retail jeweler allowed window shoppers of its store at night to light up its displays after hours, and the owner reported, "The scheme caused considerable comment."[33] Similarly, proprietors of a hat store also used this technique of push-button lighting in targeting guests of two nearby hotels, hoping that those in proximity would "remember 'that store with the push button.'"[34] In these cases and others, storeowners strove to grab potential consumers' attention and create a spectacle as well as associate push buttons with pleasure.[35] They tried to capitalize on the fact that people liked the sensation of pushing a button to create an experience like Johnson's, where a person could automatically conjure anything she desired.

Some, especially those from religious backgrounds, protested against this "electric-button theory of obtaining our desires" because they worried that an emphasis on getting such luxuries "strains the muscles and empties the purse; and there still remains in these restless souls a great yearning of unsatisfied desire."[36] Whereas advertisers and manufacturers saw the cultivation of desire as a way to achieve their economic aims, complaints like this one questioned whether material goods could ever provide a sense of true satisfaction and reflected longstanding suspicions about secular or capitalist hedonism. However, efforts to groom the desiring consumer only increased with time, and the use of push buttons as the "center of attention" in the consumption interaction provided an avenue for encouraging fingers to help themselves to that which they could afford or only dream about.[37]

To this end, interest continued in thinking through how buttons could provide any satisfaction, including by providing the kind of push-the-button information that Nesbit and Rieber had imagined years earlier. Some models existed only in theory, as in the public library of the future, where the "Automatic Who, What, and Why Machine" would feature "an attractive case of several hundred buttons" that would answer questions of nearly any subject matter.[38] Although this idea may have seemed far-fetched at the time, a number of cities rolled out experimental designs to prototype this concept. An "automatic city directory" in Miami, Florida, and a push-button map in Los Angeles, California, could assist visitors so they could push for locations they desired to visit, which would correspond to a red light showing its placement geographically (see figure 9.3).[39] Another called the "Informator," created in Berlin, Germany, featured 180 buttons and operated with a coin that would dispense a card supplying directions.[40]

These machines took a "vending" approach to information by relying on buttons as their operating mechanism. They offered a model for thinking about the ways that machines could catalog, store, and dispense knowledge with only a finger push.

Similarly, museums began installing push-button–activated exhibits that could generate information. In one instance, an exhibit regarding child welfare titled "What to Do" featured an electric wall chart. According to a description of its operation, "The spectator is instructed to 'press the button to find out'

Electric Directory of Greater New York in Use in a Prominent Hotel.

Figure 9.3
Prototypes of push-button guides on streets were designed to give users
information at a touch.
Source: "An Electric City and Street Guide," *Electrical Review and
Western Electrician* 70, no. 4 (1917): 1026. Image courtesy of the Library
of the Pennsylvania State College Department of Electrical Engineering
via Google Books.

where to go 'if you want to adopt a baby,' 'if you know a case of cruelty to children,' 'if a poor family applies to you for aid,' etc. Opposite each question is a push button which is connected with an electric light behind a transparency, on which is inscribed the name of the organization to be consulted."[41] Likewise, passengers on subways or steam trains could access "a convenient guide and directory and are saved the results of errors and the trouble of asking questions by an electric street announcer. ... When the button is pressed, a bell rings to attract attention. A legible sign slides into place and shows the principal stores, theaters or other buildings near by."[42] These prototypes of push-button information hearkened back to bell ringing and annunciators as a form of summons, where a push could indicate the item someone desired, yet they offered a potentially streamlined model of service by doing away with the attendant.

All of these mechanisms for push-button servitude required complex choreography among button pushers, electrical apparatuses, and architectural and electrical infrastructures to make the "magic" of pushing a button work. The electrical industry as a whole had stayed focused on selling idealized, simplified views of push buttons and their associated technologies to the public for more than 40 years.[43] In the words of Edison Electric Illuminating Company General Manager, William Brock, "everyday folk" need no longer have "mingled feelings of awe and distrust" because now electricity was "so pliable to man's will" and should trigger emotions of "admiration and confidence."[44] This bolstering of electric energy sought to remove fear from potential users by portraying buttons as one-touch tools that could convey unlimited power to the masterful button pusher via an unflappable electrical servant. However, as electricity and its associated mechanisms became familiar to consumers, the electrical industry faced an apathetic populace who took its services for granted; this posed a significant problem, as companies and suppliers struggled to make a case for themselves in the twentieth century.

Indeed, although members of the electrical industry proper often abstained from educating electricity users, in later years, corporations began realizing the value of allowing consumers to peer behind buttons' facades so as to demystify their magical properties. Previous efforts had promoted a view of buttons as safe, easy conduits for electricity that anyone could use and maintain; they relied on buttons to make electricity tangible and tantalizing but not dangerous. In a period of high anxiety about electricity, this strategy helped to portray push buttons as safe while bolstering the hands that pressed them with a sense of mastery and agency. The industry's umbrella organization, NELA, took up a new strategy that focused on demystifying how buttons worked. This project—far from altruistic in nature— worked to generate consumers' sympathies and "sell the public on the need of giving a larger revenue to the electric light and power companies of the country."[45]

In 1921, the NELA produced a motion picture titled "Back of the Button," a 15-minute film that demonstrated "the huge investment which makes electrical service possible."[46] A year later, it had appeared at 2,027 theatrical showings and 115 nontheatrical showings, with more than 2 million people viewing the film.[47] Additional showings in Puerto Rico, the Philippines, and Canada further served to spread the organization's gospel about the vast inner workings of electrical services. It is noteworthy that the NELA's specific language—"back of the button"—employed not only by the association but also by other electricians, further reinforced a sense of push buttons as outsides or "faces" of electricity that had previously masked what happened behind them.

Yet even in these educational pieces, the author did not make the network of technologies or people visible to the reader, so as to genuinely educate her; rather, it depicted the feminine, fetishized hand delicately pushing a button (see figure 9.4). This opaque image continued to hide any messiness from view, choosing to put forth a limited rendering of magical button pushing.

However, much as various constituencies tried to employ push buttons to put the undesirable out of range, the fact remained that buttons could only work when human beings and machines served their pushers, whether directly or indirectly. Indeed, one anecdote in a fiction account (1923) about miscommunication between a passenger and employee on a train suggests the problematically blurry boundaries between push-button electrical service and human help. According to the writer, "After several stops of the car, the lady, getting nervous, poked the motorman-conductor in the ribs with her umbrella. ... trying to conceal his anger, [he] gave the lady a sharp look and said: 'Pardon me, I am no push button.'"[48] The motorman did not wish to be literally "pushed" into action any more than a servant wished to be called into action. The poke of the umbrella drew attention to the ways that people often felt forced into action despite the seeming forcelessness of pushing buttons.

In quite a similar instance, yet this time in reference to electricity service, Preston S. Arkwright, president of the Georgia Railway and Power Company, admonished consumers for attributing the effects of their pushes to their own fingers (1922):

I hope you don't think that because you pressed that button you made that lamp glow yourself, or that you have so exaggerated an idea of your psychic power as to imagine your spirit called forth these spirits of

Figure 9.4
Electric companies continued to associate femininity and domesticity with button pushing.
Source: The New York Edison Company, "There Is More Leisure for the Housekeeper in Electrical Housekeeping," 1924. Image courtesy of the Warshaw Collection of Business Americana—Electricity, Archives Center, National Museum of American History, Smithsonian Institution.

service. Indeed, I don't want you to believe there is anything magic or mystic about it. You are able to get this service because of the genius, initiative, courage and effort of men and the investment of millions on millions of dollars to provide the physical facilities required and the constant attention and work of a multitude of human brains and hands to direct, control and apply them.[49]

Arkwright's chiding words, much like the motorman's, spoke to the enduringly problematic relationship between human and technical labor when it came to button pushing. Since the late 1800s, people had used push buttons to hide labor—from wires to workers—as much as possible to advance an ideal of digital command. Now, a user had to decide for himself "whether it is a more intelligent explanation to say that a push-button rings because of [his] thumb or because of the electric current in the wire which the pressure of his thumb has accidentally released."[50] This ambivalence exposed both the persistent pleasure and panic of button pushing. Whom or what really did the work when pushing a button? The button pusher's finger? The servant? The machine? The utility company? The electrical force? The practice of button pushing worked to make all of these social and technical elements disappear, whether relegating them to remote parts of the house, factory, or office, or plastering them behind walls and faceplates. A first-generation society of button pushers struggled to arrange everything it desired within reach of an effortless finger, while putting everything difficult, complex, socially undesirable, or dangerous out of touch. Quite often, it seemed, this mess of elements—waiting in the wings to spring into action—also liked to push back.

10   Conclusion

A nation of button-pushers, the Russians call us. Life is so simple, they sneer, when all Americans have to do is snap an electric button—and now it is even easier than that.[1]

By the 1930s, push buttons had achieved status as familiar communication and control mechanisms. Buttons' popularity related in part to their design and inexpensive construction. They could blend "flush" into walls and hide in pockets, making them attractive features when the newness of electricity and "automatic" machines threatened existing social structures and patterns. They hid wires and other "messy" aspects of electricity that could undermine harmonious, pre-electrical environments. By reducing technological interactions to two choices—push to "make" current and push to "break" current—buttons could act as the unintimidating faces of electrical experiences and help early adopters to understand how electricity worked. Yet the growing turn toward buttons over other kinds of control mechanisms stemmed from an even deeper ideological shift in expectations about what hands should do with machines. To push a button represented a particular fantasy of what I have termed *digital command*, where (certain) hands could direct anyone or anything to submit to their will. No longer did "manual" refer to effort and strain; rather, the gentle or "mere" touch of a button promised that only fingertips need engage with bells, lights, vending machines, elevators, or cameras. This "reversal of forces" that centered on hand practices—where a small human force could put great electrical forces into motion—suggested that human beings had truly tamed nature by sublimating it to a push.

Digital command encompassed not just new ideas about touch, pressure, and force, but it also represented a change in thinking about bodies and spatial arrangements. Architects, builders, and electricians, as well as manufacturers of automobiles, desks, and other consumer technologies, began designing spaces and devices that grouped controls around the operator, prioritizing reachability. In these models, push buttons should surround the user so that one might function like an armchair general inside of a kind of command center; the user could remotely spur servants or machines into action from one position. Buttons attached to hanging cords, affixed to walls above beds, sunk into furniture, and embedded in steering wheels each prioritized accessibility as an asset to the operator's comfort and safety, while also prioritizing discretion to make the mode of control unassuming, magical, or polite. Slogans such as "talk, don't walk" and "have every machine at your fingertips" related closely to movements based on efficiency, scientific management, and domestic engineering, where a person should minimize steps and maximize output. These ideas predated any formal "human factors" or "ergonomics" disciplines, but they focused similarly on creating a particular configuration of human–machine relationships that prioritized management and comfort within their environments. At the same time, pushing a button could spur action across distance, generating effects well beyond what the button pusher could see, hear, or even experience directly. As hand controls moved closer, so too could one's reach potentially extend further.

This ideal of digital command existed importantly in relation to prevailing attitudes about service and servitude. Indeed, push buttons' earliest primary usage related to calling servants, bellboys, office assistants, and so on to get attention; the act of pushing a button went part and parcel with servant culture and invoked references to being served. In this regard, the digital commander with a finger on the button might have the most comfortable environs and the gentlest touch, but the button pushing always occurred in tandem with another person's or machine's efforts to make that push possible. The effectiveness of buttons at conveying a message relied on servants of various kinds responding instantaneously and at a moment's notice, thereby conjuring the button pusher's desire out of thin air. Advertisers capitalized on this notion to sell buttons as conveyors of pleasurable and effortless consumption. As the Eastman Company's tagline for the Kodak camera promised, "You press the button, we do the rest." It was "the rest" that should remain hidden from the button pusher—made invisible and uncomplicated so that buttons seemed to operate automatically, magically, and submissively.

In actuality, laborers in a variety of fields, from factory hands to servants, suffered at the hands of button-pressing managers and well-to-do housewives. Craft workers, such as photographers and developers, worried about the fate of their jobs as mechanization increasingly impinged on their territory. Frustrated streetcar conductors and vending machine operators defended against enterprising children and pranksters. Electricians struggled to legitimize their field in the face of technologies such as electric buttons that were often viewed disparagingly by their colleagues for their simplicity.[2] In each of these cases, power relations emerged between haves and have-nots, between experts and laypersons, between managers and employees, between women and men, between the hands that gripped, sweat, and toiled and those that directed. Buttons did not necessarily create these tensions, but they certainly illuminated and sometimes magnified them. As users' finger force decreased, it seemed that demands for people to function as willing servants only increased. The problematic of the unwilling or too slow servant generated a need for machines to fill such a gap, to function as silent and unseen laborers whom could make the ideals of button pushing possible.

As this book has demonstrated, a wide chasm existed between the ways people talked about and romanticized buttons for being "simple," "mundane," and "magical" and the terms of their actual use. Buttons often malfunctioned, caused confusion and miscommunication, exacerbated conflict, and generated concern. Meanwhile, mischievous button pushers often found ways to manipulate buttons for the pleasure of pushing. Buttons' design meant that users could only signal and transmit limited information through sound or signs in a one-way form of address. Although inventors and electricians intended for this technology to streamline control and communication, much in the way that telegraph keys worked, it often irritated signal receivers, who might interpret presses as overly authoritarian, lazy, or rude. Button pushes gave little in the way of feedback to their pushers beyond simple signage that read "Push" or "Press." A press of a button could only transmit a signal and nothing more, necessitating either face-to-face communication for follow-up or elaborate annunciator systems to clarify one's wishes and needs.

As a result, usability problems often plagued novice users who made faux pas through inadvertent pushes that caused undesirable or unexpected effects. Buttons quite often failed in the estimation of people who pressed them. These limitations in what one could communicate through a button paved the way for increased uptake of devices such as telephones—which often began as push buttons for closed-circuit communication—as the twentieth century progressed.

For one particular set of users, children, push buttons especially represented a site of confusion. Sectors of society from educational institutions and the press to electricians and parents asked children to habituate themselves to the daily gestures and sensibilities necessary for pushing buttons, to imagine their bodies as powerful and capable by employing mechanisms for machine control. As children increasingly "tuned" their hands to perform this control, their acts of touch simultaneously threatened the norms of quiet, order, and hierarchical power relations, which dictated that children should only use their hands in those contexts that adults viewed as fit for their use. In this regard, the Industrial Revolution produced a struggle to combat out-of-control bodies even as so many campaigned to put bodies in control. Children served as convenient test cases for determining the physical boundaries of socially acceptable pushes. Boys, as "embryo Edisons," were taught to use their hands for experimentation, creativity, and play, but they also incurred admonishment for their "unruliness."[3] Meanwhile, gendered stereotypes reinforced the notion that the most "basic" human beings, "dainty" and "fragile" girls, would benefit from buttons' capabilities or would require protection. Disciplinary measures—locking down buttons, initiating lawsuits, or removing buttons altogether—were the most common strategies for dealing with the first generation of button pushers. Such measures often conflated the gender and character of the presser with the device's affordances, making it unclear who possessed the privilege to touch.

The historical moment described herein offers a particularly compelling snapshot because most homes did not have electricity, only a small segment of the population had access to push buttons, and push-button culture would not emerge, full-fledged, until the 1950s in the context of Cold War paranoia and space age consumerism.[4] Indeed, much more talk occurred about push buttons and the arrival of the "push-button age" in comparison with those whom actually had access to buttons and used them on a daily basis. Their marginal and experimental status meant that users necessarily had to grapple with basic questions that would somewhat fade into the background once these mechanisms achieved widespread uptake. These questions included: How should buttons work? Who should have access to them and under what conditions? What kind of user experience should the push button offer? As inventors tinkered with push-button designs, and as electrical services became more widespread, the electric industry's once hyperbolic promises could be more thoroughly realized in practice to some extent. People no longer peered behind buttons to see how they connected to a network of wires, bulbs, outlets, batteries, generators, or central stations. Most of the time, users would come to push buttons in ways that masked both complexity and power relations behind their pushes.[5]

Following World War II, discussions and plans for push-button living reached fever pitch. In fact, of any historical moment, the 1950s might most be labeled the "push-button era," when promises of idyllic suburban American consumerism and Cold War concerns trickled down into every kind of consumer product, and when television shows such as *The Jetsons* envisioned a future of flying cars, robots, and push-button solutions to all manner of problems. Because buttons tended to symbolize limited, proscribed kinds of interactions with machines that made technology appear effortless and automatic, individuals and organizations continued to rhetorically construct these surfaces as either heroes or villains, with little nuanced discussion. On the positivist side, push buttons appeared as the centerpieces of "technological fantasias," in which "the push of a button replaced the wave of a magic wand in fairy tales as a tool to accomplish the unlikely."[6] Hearkening back to the "invisible genie" often promoted by the electrical industry in the nineteenth century, prognosticators called on push-button "automation" as the wave of the future. Once again, an emphasis on fingers figured prominently in these discourses. Reporter Alfred Leech proclaimed in 1956, "Your New Home of 1980 May Operate by Thumb," thereby making the United States "the land of the big thumb."[7] Advertisements for drive-thru restaurants, jukeboxes, Wurlitzers, television remote controls, and push-button automobiles, to name a few, displayed prominent images of fingers on buttons to demonstrate the ease of digital command imagined by their predecessors, with taglines such as, "Push 'N Dine, A Complete Dinner ... at your fingertips!"[8] These continual proclamations fell in line with broader appeals to technology as a solution to hard labor, but their specific focus on hands and reachability is notable.[9] Romanticized notions of button pushing persisted not only because they helped corporations and inventors to sell products, but also because they gestured toward a better, high-tech future of "perfect" control—at a touch. They married technological solutions to labor with a particular way of using one's hands to direct and manage via complex, "modern" control panels rather than by toiling.

On the whole, the "push-button age" of the mid-twentieth century reflected many of those aspirations and fears present more than 50 years prior. Yet where the late nineteenth- and early twentieth-century public had only begun to experiment with self-service, automaticity, and push-button control, the turbulent social and political climate of the 1950s and 1960s ushered in a full-fledged effort to bring these concepts to fruition. This drive to "buttonize" every aspect of life emerged directly out of an automation and efficiency craze that sought to reconfigure how humans related and delegated to machines. What if everyone could live in suburban prosperity, controlling their environments much like a pilot in an airplane or a space shuttle? What if the tasks of everyday life could be made faster, more instantaneous, and more reliant on machines? Where one or two push buttons to control lights, an elevator, or a bell would have sufficed in previous years, automobile dashboards and high-tech kitchens began to resemble complex command centers; the more buttons, the better. An advertisement for the 1959 De Soto automobile proclaimed it was "The Car That Has All Its Buttons," and the company boasted, "Almost everything in De Soto works with push buttons. There's push-button drive, push-button heat and push-button entertainment. Also available are push-button power windows—and push-button power seats that adjust six ways to fit you perfectly."[10] An abundance of buttons functioned simultaneously to bring about comfort and fend off ever-present anxiety about a potential nuclear future.

As in the past, advocates for push-button products and services viewed women as their primary demographic, assuring the housewife that she could, in the words of J. W. Alsdorf, president of the National Housewares Manufacturers Association (1950), "sit back and take her ease while the work is done for her."[11] Not only did prophecies like Alsdorf's greatly exaggerate the availability, affordability, and efficacy of push-button living for most women at this time period, but they also spurred outrage on the part of some for how they portrayed women's interactions with technology. So argued Jessie Cartwright, in an opinion piece that scathingly denounced advertisers' treatment of women button pushers:

Have you ever been guilty in your ads and brochures of telling [a woman] that she doesn't need to know a thing? 'Just push a button, and put that pretty little hat on your saucy little empty head and take in a double feature movie!' ... Now, I ask you! How would you like your power drill or your amateur bench lathe LET you operate IT—and assume that you had no will or brains?[12]

Cartwright particularly took issue with how advertisers envisioned women as passive, thoughtless operators of machines who "let" machines do the work. Instead, she called for a view of women as competent and capable of learning a machine's nuances, dictating what it could do and how it worked, much as some had argued in decades past against the concept of "foolproof" machines. Yet stereotypes of women as ideal button pushers persisted, particularly in advertising media—implicating women as fragile and technologically incompetent much as they did in the nineteenth century—as a strategy for promoting buttons as antidotes to the drudgery of housework and promoters of domestic pleasure.

Lasting innovations such as push-button telephones also came to fruition in the early 1960s. Debuting at the Seattle World's Fair in 1962, Bell Telephone's button or "touch-tone" phone featured a configuration of buttons that replaced the common dial after extensive human factors testing. The company sought to demonstrate that its scientific process created a justification for switching to buttons, emphasizing how psychology, design, and ergonomics came together to produce the ideal button arrangement.[13] Documentary footage from the fair portrayed teenagers comparing dial mechanisms of yore to push buttons in an interactive display, with the demonstrator encouraging the button pusher that she could see "how many seconds you save the new way."[14] Once these phones hit the market, further advertisements put emphasis on speed, simplicity, and fingers, displaying the all-too-familiar extended pointer finger (usually a woman's) as evidence of yet another step closer to a world entirely controlled by buttons. Talk of "touch" as a metaphor and a hand practice figured prominently in ads disseminated over the years. To "keep in touch" or "get in touch" often referred to telephone communication.

Other prominent push-button technologies such as television remote controls also gained traction during this time period, with the Zenith Radio Corporation taking the lead on early prototypes such as the "Lazy Bones" (1950), "Flash-Matic" (1955), and "Space Command" (1956) remotes. The names and branding of these devices referred to familiar facets of digital command that emphasized comfort, command and control, touch, and reachability. Indeed, an ad for "Lazy Bones" promised, "WHY You Can Operate Zenith TV from Your Easy Chair." The operator need only press "lightly with your thumb" due to a "miracle of automatic precision and stability!"[15] Much like the button pusher of the nineteenth century summoning servants from a dinner table or honking a horn at the steering wheel, the television viewer of the twentieth century was encouraged not to fumble, tune, or even get up from their easy chair. As with many inventions of this time period, television remotes prompted ambivalent responses from consumers, torn between the ideals of effortless consumption and concerns over automaticity, agency, and laziness.

Critiques of push buttons in the mid-twentieth century took many forms, and observers continued to fear, as in the past, that the act of pushing buttons would erode human beings' morals and work ethic and, in the worst-case scenario, destroy the planet in a single instant. By 1956, at the height of the Cold War, journalists told readers that although a push-button warfare scenario "sounds like a science fiction nightmare," in fact it was "frighteningly real."[16] Politicians, reporters, filmmakers, and scientists injected the push-button icon with their greatest anxieties at this time period, littering cultural products from the film *Dr. Strangelove* to the pages of magazines with suggestions about the nation's fate in a world controlled by buttons. Narratives in these forums worried about buttons' imagined irrevocable nature, which dictated that, once pressed, the button presser could not turn back. Most stories about push-button warfare focused on conjecture and speculation, forecasting a future that never arrived; some called push-button warfare a "myth" and the purview of "calamity howlers," whereas others sought to identify the exact day that inevitable war-by-button would arrive.[17] The panic over such visions led to prominent figures such as architect Frank Lloyd Wright (1946) to bleakly conclude, "The push-button civilization over which we were gloating has suddenly become a terror."[18] Ethical quandaries about button pushing revealed that the way buttons presented choices to human beings—as "all or nothing" or cause and effect—could take a real psychological toll. In fact, social psychologist Stanley Milgram's famous experiment (1961) exposed this problem, as he investigated what would happen if he could convince participants to administer electric shocks to other study subjects via a push button with labels such as "Danger: Severe Shock."[19] He demonstrated that buttons functioned as psychologically seductive tools—people wanted to push them, especially when distanced from the consequences of their pushes, and they exhibited a great deal of obedience in following Milgram's requests. Although pushers subsequently experienced anxiety and guilt about what they could accomplish remotely with a button, the study's results proved worrisome in that one could easily manipulate a button pusher to carry out brutal and morally repugnant acts.[20] Such findings carried particular weight in the context of Cold War anxiety and in the aftermath of Nazism.

These kinds of experiments pointed to the fact that as much as push-button consumerism in homes represented a welcome "space age," postwar mentality to daily life, it also invoked deep fears about the moral and political implications of pushing buttons. Now, it wasn't the bumbling, ignorant gentleman in a top hat of the 1890s who accidentally blew up the world, but rather a maniacal Russian or desperate American who planned to carry out nuclear destruction with a finger. As the U.S. military contemplated increasingly automated solutions to warfare, the image of the button-pushing soldier reflected a fear of automation and masculinity gone soft. In the words of reporter Jack Geyer (1952), "I conceive the push-button aviator as pale, wearing large glasses and slightly round-shouldered," and he believed this pilot stood in contrast to the heroic one of former days who, battling the elements and his enemies with his bare hands, had "20/20 vision and a slightly dashing manner."[21] Whereas feminine button pushers were perceived as empowered consumers, finally relieved from their physical suffering, emasculated male button pushers sat lazily reclined with pillows and watched war transpire on TV. Even U.S. President John F. Kennedy went so far as to express concern that Americans were becoming "Pushbutton Softies," recalling long-held stereotypes of digital commanders as nonworkers and button pushing as nonwork.[22]

Due to the concerns of that historical moment, buttons achieved larger-than-life notoriety in journalistic accounts, advertisements, and popular culture narratives. Both utopian and dystopian visions of a "push-button world" for all far exceeded how users employed push buttons on a daily basis, and yet an examination of technologies of the mid-twentieth century reveals an incredible investment in making every control mechanism a push button. For those accustomed to city life, in particular, the act of pushing buttons became increasingly familiar yet unsettling in relation to concerns over automaticity: "What modern city dweller hasn't felt a twinge of fear at entrusting himself to a pilotless elevator, a fast-moving escalator or some other contraption which wrenches his fate out of his own hands?"[23] Push-button traffic signals at crosswalks represented one such innovation that provoked anxiety, with concerns about safety that manifested similarly with regard to elevators. The concept of self-service, tested out with early vending machines and elevators in the late 1800s, took root in these devices to make consumers more responsible for managing their transportation, consumption, and communication without human attendants. Yet many worried that by giving over much of their responsibility to machines, the push of a button represented loss of control rather than gain.

To consider how push-button mechanisms have achieved such an entrenched position in the practices of everyday life, it is critical to consider how designers began to think about buttons as part of computer technologies beginning in the late 1960s. Although this book has treated "digital" operation as a function of the digit—the finger—such a term inevitably recalls the digitalness (consisting of binary 1s and 0s) of computers. Buttons and computation were first significantly linked together in 1968 when Douglas Engelbart created the first prototype for a computer mouse. After eliminating other possible designs to manipulate digital items on computer screens like light pens and light guns, Engelbart settled on the mouse as a suitable concept. He commented about the invention that, "You didn't have to pick it up and you could put buttons on it, which helped."[24] It is notable that the inventor viewed the mouse, according to one author, as "part of an effort to optimize basic human capabilities in synergy with ergonomically and cognitively more efficiently designed artifacts."[25] Like his predecessors thinking about automobile steering wheels with buttons and electricians considering the placement of push-button light switches, Engelbart concerned himself with how humans could more naturally and strategically engage with computers through their hands as well as their whole bodies as part of systems for communication and control. In this particular instance, Engelbart sought a solution for how computer users could multitask by using their hands in tandem with, but independently from, a screen. The "clicks" and "clicking" behavior that would become hallmarks of personal computer usage and web surfing began here, profoundly tied to pushing buttons. Indeed, according to Logitech in an early advertisement, the mouse served as "a hand's best friend" with "buttons for maximum comfort and minimum fatigue."[26] Hearkening back to prototypical and imaginative designs like those for the "Informator" or the "Automatic Who,

What, and Why Machine" of the early twentieth century, computer mice structured people's interactions with information by using a long-held model of placing buttons at one's fingertips. The mouse user as digital commander embodied all of the values of digital command—tethered to computer hardware so that the user might sit, direct, and manage with clicks of the finger to make the computer do the user's bidding—while the wires, mainframe, and other hardware remained safely enclosed and removed from the operator.

Meanwhile, a shift in computing occurred from a text-based, command interface to one of graphical user interfaces (GUIs) that permitted computer scientists and designers to incorporate icons and images into the end user's experience. Although a history of GUIs is too lengthy to recount here, it is of particular note that buttons began appearing on desktops, and later on the first Web browsers, as "a way of dressing [the computer interface] up as something well known and well understood," while demonstrating users' "limited understanding of the computer as a machine and as a medium and how it functions in culture and society."[27] Buttons accompanied other familiar icons such as manila file folders, garbage cans, windows, and floppy disks that mapped physical experiences onto digital ones to produce unthreatening and appealing metaphors for computation.[28] Before long, button designs had almost entirely replaced textual hyperlinks on the Web, with so many buttons populating the browsing experience that some designers in the early 1990s called for "ending the tyranny of the button," which seemed "unlikely to happen whilst hypermedia authors and users have such a fixation on buttons."[29] Although designs ebbed and flowed over time, vacillating between overt, brightly colored icons and starker layouts, buttons have persisted across every area of computation, creating a kind of "'buttonization' of culture in which our reality becomes clickable."[30] Theorists have worried to what extent these buttons have produced false agency and even a dangerous user experience in digital contexts, at once positioning users as masterful and in control of their decision making while also constricting their choices to binary ones when they cannot fully understand the implications of choosing to "accept" terms and conditions or "buy" a product.[31]

It is unsurprising that evangelists and naysayers alike have historically taken hold of push buttons (and continue to do so) as an emblem of life in technologically saturated environments because machines of all kinds generally tend to grip the public's imagination and generate larger-than-life fantasies and fears.[32] People continue to worry about laziness. They reprimand "armchair generals" who push a button that sends a drone across the world. They express concern that button pushing either prioritizes or jeopardizes safety (as in recent conversations about push-button starters for cars). They voice frustration that children press buttons constantly. In some ways, then, it seems that nothing has changed at all from more than 100 years ago, even though the act of pushing buttons is more thoroughly entrenched in everyday life. Issues of distance and human contact, automation and human engagement, laziness and effort demonstrate that buttons signify the best and worst of technological solutions to human problems.

Yet, importantly, there is nothing natural or inevitable about buttons or the act of pushing a button. Various constituencies over the years—especially advertisers and manufacturers—have marshalled tremendous resources to make buttons popular and alluring. Sometimes these efforts succeed, but quite often they fail; the case of push-button light switches, for example, demonstrated that buttons weren't the best long-term technical solution. Similarly, the rarity of push-button bells in other countries to summon servants indicated that button pushing was (and remains) a culturally and socially situated practice, deemed desirable or unattractive depending on context. As Foucault once remarked (1980), "One needs to study what kind of body that the current society needs."[33] As this book has suggested, the button-pushing body emerged in the late nineteenth century, but far from one coherent body, there existed many kinds of button pushers well beyond the imagined ideal body. At the same time, these bodies stood in tension with other kinds of bodies, especially those perceived as laboring ones.

It is important, then, to continue to track how perceptions of button pushers manifest at different historical moments. In particular, they have made great fodder for all kinds of commercial and pop culture products in the twentieth and twenty-first centuries, continuing a trend begun by Edison Illuminating Company and others that playfully advertised pushes through poems, stories, and jingles. Songs such as "Push a Little Button" by Ninette (1966), "Push the Button" by Money Mark (1998), and "Push the Button" by Sugababes (2005) all play on common images of buttons as binary, automatic, high-tech, and sexually pleasurable, and reflect the moments in which they were produced.[34] Similarly, television shows and films such as *The Jetsons* (1962–1963 and 1985–1987), *The Outer Limits* (1997), *Lost* (2004–2010), and *The Box* (2009, based on the 1970 short story "Button, Button") examine science fiction, moral issues, and what-if scenarios about the push button as forbidden fruit or as object of catastrophic destruction.[35] More recently on reality TV, too, shows such as *The Voice*, a singing competition, make large red buttons the centerpiece of their drama; famous singers acting as coaches sit with their backs turned to hundreds of hopeful performers as they audition. A touch of one's button from a large, comfortable chair signifies acceptance onto the program and therefore a possibility of achieving rags-to-riches success. Notably, only celebrities have access to the button; they control the fate of contestants vying to demonstrate their worthiness. The show sets up a tension around pushing or not pushing as a strategy to embolden the judges as gatekeepers, and they exemplify the quintessential digital commander who manages and orders with comfort from a position of privilege.

Corporations, too, have clung steadfast to producing push-button products, marketing not only devices with push buttons—but also buttons themselves—as gratifying, collectible objects. For example, in 2005, the Staples Corporation began its "Easy Button" campaign to celebrate the simplicity of Staples' office services, first with advertisements and then by selling physical red buttons (around the size of a paperweight) with the word "easy" printed on them to encourage individuals to place them on their desks. Upon pushing, the buttons would return a recorded male voice stating, "That was easy." More than 1.5 million customers had purchased a $4.99 button by the end of 2006, with the campaign earning praise as a "marketer's dream."[36] Of particular note, users often purchased and "hacked" these buttons to transform them into something new: for prop comedy, Secret Santa office rituals, and even placing them humorously in airport cockpits.[37] Some of these do-it-yourself projects responded in opposition to the trope of "easiness" espoused by Staples, as in the case of industrial designer Al Cohen's online diatribe against the campaign, where he encouraged blog readers to repurpose the device as an "Evil Button."[38] Cohen stated a number of rationales for transforming these buttons, chief among which included his statement: "As a business owner of 25 years, I found the mere notion of pushing a magic button which solves all problems a slap in the face. Business is not easy. ... I couldn't sit idly by when I saw this."[39] Cohen's comments speak to users' enduring problematic relationship with doing things the "easy" way, much as marketers tout convenience as the driving force of consumerism. More broadly, the negotiations that occurred between Staples and consumers over the "Easy" button demonstrate the continuing iconicity of buttons as symbols for simplicity, as well as how users continue to experiment, play with, and shape buttons in ways that defy and exceed their original purpose.

In perhaps the most striking example of push-button culture in the present day, we might consider what transpired on April 1, 2015, when Reddit, an entertainment, news, and social media website, launched an "April Fools'" challenge that quickly snowballed into much more. At a surface level, Reddit's homepage looked outrageously simple. Titled "the button," the screen featured a red circle and a countdown clock. The premise: any Reddit user, registered before April 1, 2015, could press or click this digital button one time; if no one in the community pressed within 60 seconds, the button would disappear for good. Reddit enticed its users with a provocative phrase, noting, "We can't tell you what to do from here on out. The choice is yours." Users overwhelmingly chose to press, with more than 1 million "pushes" registered in a little over two months. The community exploded with factions of pressers and nonpressers, each deliberating the merits of the button and what might happen when the clock ran out.

As all of this activity transpired online, journalists, pundits, and psychologists jumped in to explain why an odd experiment had attracted so much attention. Some offered explanations about addiction and a fear of mortality. The Brookings Institute proposed that "the button" provided useful lessons for improved public policy and crowd-sourced government. News articles and blogs called the act of pressing the button "fascinating," "ridiculous," "craze-inducing," "bizarre," "hypnotic, divisive and possibly evil." Yet far from outlandish or new, the experiment recycled a familiar image of buttons as simultaneously pleasurable and panic inducing that has perpetuated for more than 125 years. As cultural and technical surfaces, Reddit's button seemed to beckon and even instruct the hand to push, offering swift and hedonistic effects. Still, it also sent a warning message due to its ambiguity and shockingly red color—like the fire alarm or warfare button: Do not touch. Hands off. Danger. With a single, irrevocable push, you may take a life or blow up the world. It's no wonder, then, that flocks of participants lined up at the altar of Reddit's red button, ready to enact their dual positions as playful consumers and doomsday harbingers.

Reddit's success could be attributed to its reliance on a popular culture formula that has graced the plotlines of so many films, television shows, and science fiction accounts. However, further unpacking the dynamics that played out over those two months reveals that this microcosm can begin to tell more about the seemingly ordinary and ubiquitous mechanism called a "button." First, for more than any other reason, people pressed the button—or didn't—as a way to demonstrate their status and standing in the community. In deliberations about pushing and nonpushing, and about how fast you could push to earn special "flair" (colorful denotations attached to one's online profile), emerged a predominant theme expressed by one user: "Pushing the button is a privilege. Don't waste it."[40] Indeed, Reddit's experiment dramatized that who could push the button mattered as much, if not more, than what effect the button would produce. Button pushers formed communities within the community according to the speed of their button pushing. "The Followers of the Shade" group consisted of Reddit members who refused to press the button. Meanwhile, "The Redguard" would only push close to when the 60-second counter ran out.[41] These communities created anthems, rules, and rituals unique to their button-pushing ethos, thereby revealing how varying social statuses could attach to the button pusher.

Second, far from a stable and unchanging object, "the button" crashed more than once, preventing users from pushing it for hours or even a day. On these occasions, commenters rushed to the site wondering whether the experiment had ended or whether the button would reappear. Once it did—rebooted by Reddit's programmers—a new set of conversations emerged around "The Great Button Crash." Now the community had to grapple with instability and fragility; with each glitch, Reddit's button lost a bit more of its magic. These two brief examples demonstrate that behind the symbolic button and its magical effects have always lurked another button and another kind of button pusher. This button behaves unexpectedly and often thwarts its user, failing to produce any effect at all. Similarly, the button pusher pushes not primarily to achieve a task, but rather to assert influence or demonstrate authority within the dynamics of a larger community. Emboldened by the right to push, the pusher gets to make a choice often denied to others.

These negotiations, not unique to Reddit's "button," animate all of the buttons that surround us each day. Company after company promises a push-button experience that makes consumption effortless and unfettered, such as Amazon's Dash

Button, with the tagline "Place it. Press it. Get it." Amazon's more than 100 Internet-connected physical buttons, made to affix to any surface in one's household, instantly order common products from detergent to toilet paper at a push. As *Wired* magazine author Klint Finley suggests in reference to Amazon's ever-expanding repertoire of buttons, "The real world just got a little more push-button."[42] Web designers and Internet companies have taken a similar approach to buttons, offering the possibility of an instantly gratifying experience now with taps and touches instead of pushes. According to technology entrepreneur David Sacks, "Today a single tap or swipe gets you a date, some flowers, a car, a movie, a restaurant, even a hotel. After that it'll get you a job, an apartment, a wedding, even a dog. It goes on. This is your life, and you're living it one tap at a time."[43] Sacks' comment strikes a familiar chord, which imagines push buttons as the purveyors of every answer, every whim, and every longing. At present, the prevalence of buttons in a highly technological world makes a desire for these whims appear not only justifiable but also ordinary. Buttons unsurprisingly fade into the background of human–machine interactions unless the mechanism in question happens to malfunction; the attractive, glowing computer power button belies the intricate chip behind it, and the calibrated radio dial obscures the electrical waves that deliver music to one's ear. Most individuals, if quizzed on how these processes work, could hardly provide an explanation—they interact with the surface rather than the machines, mechanisms, and people that power them.

Yet users' relationships to digital and physical buttons, so ingrained in the fabric of everyday life, rarely transpire so unproblematically. From the press that turns on a television to the push that sends a text message or drone halfway around the world, pushing a button (or today tapping it) necessarily involves power relations and politics. It matters who gets to push the button (and who does not), just as it matters that button pushing necessarily puts someone or something else to work in its stead. Every push—every desire gratified—requires a mass of human and technical labor to make it possible. The myth of the "invisible genie," or the belief that a tap, push, or touch can make anything possible, dangerously ignores the fact that, to recall one Reddit user's words, pushing buttons is a privilege—a special benefit or right only available to some. Toilet paper, wine, a cab ride, and information do not appear out of thin air, much as button pushers are encouraged to think in these terms and to imagine themselves as "digital commanders" who get to command and control from the comfort of their chairs, smartphones, and laptops. In this regard, digital command can have destructive implications, with ever heightening expectations on the part of consumers and digital providers that buttons should give users what they want, when they want it, and wherever they want it, all the time. Although talk of genies and invisible servants no longer appears front and center, the implication of button pushers as masters dictating from on high remains the same.

Although much has stayed consistent over more than 100 years of history, it is clear that much has changed both for button pushers and buttons. From a technical perspective, push buttons currently perform so many functions now (especially as digital rather than mechanical buttons) that they don't always work in line with users' expectations. Some buttons offer a number of choices by opening up menus when the user holds a finger down on the button for an extended period of time, whereas others start a preprogrammed response that happens entirely outside of the user's control. These multiple states and functionalities create less clarity about what a button does because many buttons no longer provide a binary condition—on/off, start/stop, make/break—and instead offer users a bevy of choices that may not be visible at the original push. Of course, "hidden button states are not ideal," as observers have noted, because they "add to the cognitive load of users and are vulnerable to operation errors."[44] Yet to call present-day virtual buttons inherently more confusing or complex than their mechanical predecessors at the turn of the twentieth century would overlook the complexities that have always accompanied buttons; the fallacy lies in assuming that buttons have ever provided a seamless experience for users. The "uncertainties of the modern button pusher" have a long legacy, in part, because designers, inventors, and marketers refuse to acknowledge the typical complexities—which are usually a default condition—of pushing buttons.[45]

Additionally, as alluded to in the previous section by David Sacks, the rampant use of touchscreens has increasingly replaced pushing with tapping and touching, with many declaring a demise of analog buttons in favor of slick, flat glass.[46] This slickness has prompted concerns among a vocal contingent of user interaction specialists and users alike about the flattening of interfaces where force and feedback matter little, if at all, as buttons that used to stick out from their surroundings become seemingly intangible. Attempts in recent years have sought to address this problem through various "haptic" interventions, such as Apple's "Force Touch" (later renamed "3D Touch" and part of the "Taptic Engine"), which aims to create a future in which users feel like "you're depressing a mechanical button, when you're really just mashing your finger against a stationary piece of glass."[47] This concept of reinjecting force back into touch constitutes not only a technical problem but also a social one related to hand–machine relationships.

Many have argued that, "humans like things that respond to touch," calling for a return to physical controls instead of the "complete and abject failure" of screens.[48] Backlash against touchscreens often takes on an emotional character. Indeed, one blogger has promised that unless touchscreens begin to require some force and feeling from fingers, "Someday soon, we're going to rage-poke a hole through some indecipherable, unintuitive touchscreen."[49] These calls to restore forcefulness to touch-screens or bring back physical buttons altogether echo conver-sations begun when producers and users first imagined a push button and the finger that would push it. In early negotiations about buttons and their role in reducing effort to a mere touch, concerns arose about how much buttons should stick out, how hard fingers should press, and what constituted a legitimate and authentic interaction between hand and machine. A utopian vision of touch without force was met by vitriolic complaints that a "mania for simplification" had overtaken the nation; people argued that pushes should reflect human beings' indi-vidual character and the fact that fingers were meant to exert some force.[50] Today, touchscreen buttons reflect yet another attempt to reduce hands' burdens and make buttons aestheti-cally "flush" with their surroundings, with many lauding the benefits of "flat" design.

However, designers and usability experts have begun to slowly acknowledge users' difficulties with these buttons, with some vocal observers contesting the definition of a "push button" in the digital age. They suggest moving away from button designs altogether, which they view as "sadly inflexible" and "not natu-ral" because they restrict users' interactions to two conditions or spur more confusion than desired.[51] Some have even proclaimed a "death" of push buttons on the near horizon.[52] Apple's Steve Jobs famously derided physical buttons, with the popular press once declaring that Jobs' trademark button-less turtlenecks and iPhones and mice free of hardware buttons stemmed from an alleged "pathological fear and loathing of buttons."[53] Jobs once described his distaste for mechanical buttons as having to do with a psychological discomfort about the on-off binary that buttons present. Confronting his own mortality, he commented in a *60 Minutes* interview that,

I find myself believing a bit more. I kind of—maybe it's 'cause I want to believe in an afterlife. That when you die, it doesn't just all disappear. The wisdom you've accumulated. Somehow it lives on, but sometimes I think it's just like an on-off switch. Click and you're gone. ... And that's why I don't like putting on-off switches on Apple devices.[54]

Recalling Thomas Edison's notion of the "snap" that could end a life via electrical execution, Jobs made a similar connection to the physical switch as a metaphor for life and death. This ambivalence about the logic of switches demonstrates a shift over time toward valuing technologies that are "always on" and ever present rather than those that permit OFF at all. Science fiction writer Philip K. Dick once noted this development, too, where, "'On' more and more predominates over 'off'; it seeks 'on.' ... 'Off' could be regarded as a lack, a failure, a defeat, an impediment to be overcome. We still have a binary system, but priority (plus value) is given to 'on.'"[55] Moving away from physical OFF buttons toward always-on technologies might demonstrate this increasing priority for ON, for doing away with the psychological and philosophical uncertainties of OFF as a state of disconnection and even death.

More broadly, in imagining a button-free future, user interaction specialists have called for popularizing other kinds of interactive experiences—from Google Glass (glasses that display the web in one's line of sight) and Apple's Siri (voice-activated software made to simulate a virtual servant) to Microsoft's Kinect (a screen-based entertainment system that relies on hand gestures and other body movements).[56] The creators of these technologies promise effortless interfacing that purports to move beyond touch altogether by using tools and sensors that respond "smartly" and automatically.[57] These technologies remove buttons from the equation and promise a transformation of one's body into a device or joystick with (potentially) endless degrees of motion, an experience that contrasts greatly with the characteristics of digital command that include "mere" touch (reduced force), reachability, sedentarism, and automaticity.[58] They are sold as the next evolutionary step in "organic" interfacing, where touch and buttons serve as limitations or liabilities. In such models, a blink of an eye or a wave of a hand could potentially eliminate all of the imperfections and frustrations caused by buttons. However, interestingly, most of these devices require the user to push buttons for ON/OFF and swipe to navigate. Even virtual reality technologies such as Oculus, which espouse touch-free immersion, require physical controls for many activities.[59] To this end, these reenvisioned systems do not eliminate tactility or even button pushing. Rather, they (often unsuccessfully) ask the user to navigate through a technological world while pretending that these control mechanisms don't exist, and instead the lines between "real" and "virtual" have blended away entirely. As a result, users find themselves making all kinds of physical and social adjustments to this universe, all the while being told that they should have a seamless interaction that doesn't necessitate any accommodations. The rhetoric that accompanies these button-less technologies often harmfully suggests that users should naturally know what to do with them—that they provide more embodied experiences than button pushing; this kind of insistency perpetuates the problem of assuming that technology should "just" work and that has attached to the logic behind many "convenience technologies."[60]

Beginning to think about what comes next, then, requires neither doing away with buttons writ large nor embracing them entirely as the promise of a yet-to-be-realized future. Rather, a constructive approach to physical and digital mechanisms for communication and control would involve recognition that buttons, screens, dials, levers, and switches exist within a complex web of social and technical interactions. No one solution will provide a magic bandage to the problem of meaningful and intuitive interactivity. However, it's important to begin thinking about the historical assumptions that underpin buttons and their counterparts to consider why certain habits, assumptions, and affordances perpetuate.

A historical study of buttons suggests that, although usability issues, technical confusion, and mischief have always plagued buttons, it's not buttons that provoke worries but rather how push buttons get embroiled in social negotiations and power relations. The most vocal concerns about buttons manifest in relation to people whose fingers on the button occupy positions of privilege and that they abuse those privileges—by using buttons unethically, by sending others to do their work, by taking lives without thinking about it, and by commanding from a relaxed position while others sweat and toil.

Two recent, widely publicized and debated examples starkly highlight these issues. The first occurred after sexual harassment and misconduct allegations were leveled at Matt Lauer, the popular host of NBC's *Today Show*, in November 2017. Reports surfaced suggesting that Lauer had used a secret button underneath his desk to lock his office door in order to "welcome female employees and initiate inappropriate contact while knowing nobody could walk in on him."[61] The public was outraged that Lauer had taken advantage of his status in the organization and pushed a button to maintain and exert his power. *Late Night* host and comedian Seth Meyers later joked on his show, "Let me address anyone who works in the button-installing business, if I may: nobody wants a button under their desk for a nonevil reason! ... If someone asks you to install a button under their desk, just nod and then report it to the police."[62] As it turned out, General Electric (NBC's former parent company) had installed these buttons years earlier in some offices for executives to make private phone calls or in case of emergency—but the potency of Lauer's push-button actions remained the same.[63] The incident served to add more fuel to a broader societal discussion about harassment, privilege, and workplace gender politics already underway in Hollywood, highlighting how prominent men could "abuse" buttons and the people made to heed their call.

The second instance, in January 2018, occurred when US President Donald Trump posted a message on Twitter regarding his capacity to engage in nuclear war with North Korea. Trump wrote:

North Korean Leader Kim Jong Un just stated that the "Nuclear Button is on his desk at all times." Will someone from his depleted and food starved regime please inform him that I too have a Nuclear Button, but it is a much bigger & more powerful one than his, and my Button works![64]

Following Trump's show of bravado, news outlets quickly reported that "the only red button on the president's desk is actually one that summons the White House steward with a Coca-Cola (really)."[65] This remark stood at odds with the potent myth of the presidential button that has persisted since the Cold

War. Trump's tweet, however, not only played upon this vivid imagery of the imaginary button but also displayed thinly veiled sexual one-upmanship—the "mine's bigger (and more virile) than yours" approach that is typically more common in school-yard bullying than in geopolitical affairs.

Both the Lauer and Trump examples demonstrate long-held societal concerns about men in positions of power, sitting behind their desks, carrying out dangerous and irreparable actions with a simple push. In Lauer's case, they illuminate micropolitics that often occur in workplace contexts, where bosses and managers take advantage of their employees by commanding from on high. Taking a step further than the nineteenth-century complaints of lazy managers, however, Lauer's purported actions suggested harm (or even evil, according to Meyers) if the button fell into the wrong hands. This concern about whose hands should have access to the button likewise followed Trump on the campaign trail before his election, with some deeming him unfit to lead for this very reason.[66] His comments on Twitter further inflamed these fears, leading to press accounts with questions such as "Can Donald Trump and Kim Jong-un simply reach across their desks and bring on Armageddon?"[67] Although reporters debunked the myth of the nuclear button following the tweet, it is remarkable that these imaginary buttons continue to play a prominent role not only in the public imagination but also in global matters with real political stakes. While the president may not have a button on his desk to trigger nuclear war, written threats about such capabilities can carry tremendous rhetorical weight. Both Lauer's and Trump's actions demonstrate that buttons take on significant meaning when situated in a particular context. If the same button can place a private call, request a Coca-Cola, and trigger nuclear war, then the button pusher's character and intentions become incredibly important.

It is clear that pushing a button is anything but "easy." Buttons both reflect and shape conceptions of gender, race, and class at a given historical moment. They continually relate to understandings of labor, effort, and control. And they reflect uncertainty about what it means to use our hands in a technological age—to the ideals and limitations of digital command. Buttons' past and present urge us to begin thinking about button pushing in relation to ethics—not only to the ways that buttons are technically configured but also to how they are used in everyday practices and deployed in advertising and popular culture. Simply put, pushing never occurs without politics. Thus, to imagine an ethical button or button pusher requires acknowledging that, although the act of pushing may require less and less physical force over time, it will continue to imply forcefulness without further consideration of the ways that various kinds of hands and machines interact. We must acknowledge that when people push, someone or something must always do the finger's bidding to make the magic of buttons possible.

Further Reading

It is instructive to consider how this book intersects with previously written histories that have covered many of the same themes and events, albeit from different perspectives. At the turn of the twentieth century, users (ranging from housewives to electricians) especially contested power—both literal and figurative—over machines within the context of the Industrial Revolution.[1] During this "crisis of control," large-scale systems such as factories and post offices and large-scale technologies such as the telegraph, railroad, and automobile served as responses to new problems posed by a revolution in manufacturing, transportation, and communication.[2] Efforts toward electrification also figured importantly, and historians have provided numerous rich investigations of developments in electrical infrastructures and societal reactions to electricity in the late nineteenth and early twentieth centuries. These accounts demonstrate that experts used electricity to distinguish themselves from laypersons, creating an elite culture with professional standards and defined electrical language.[3] Electrical systems of power were indeed complex systems made up of a variety of people, institutions, and technologies that necessarily had to negotiate both political and technical matters.[4] Additionally, electrical development was

socially shaped and contextualized, taking on a shifting character in domestic spaces, in towns, and at world's fairs, with many
often spiritualizing and romanticizing electricity's capabilities,
and others experiencing a great deal of anxiety about electrical
dangers and ills.[5]

Despite these texts answering many questions about electrification, some still remain. Previous studies of electricity are
by and large immaterial—they are void of the mechanisms of
everyday life that facilitated (and hindered) access to electricity.
Indeed, what is a light bulb without the switch that powers it
on? What is the spectacle of light without considering the other
ways that people experienced electricity with their bodies—for
example, what did it feel like to "touch" electricity with one's
fingers or through a conduit such as a button? To what extent
did the act of pushing a button serve as a gateway for early users
to better understand how electricity worked? How did different
kinds of materials, from ivory to brass to rubber, take on different aesthetic connotations for electrical interactions? Without getting at these details, we lose sight of the ways that users
controlled electricity in a host of environments.[6] To accomplish
such a task requires analyzing macrolevel structures, widespread
technologies, and institutions where electrification took root
(e.g., factories, offices, bureaucratic measures, the telegraph) as
well as paying attention to "mundane" places (including homes,
streetcars, automobiles, vending machines, and elevators) and
practices (ringing a doorbell, honking a horn, etc.).

In tandem with a movement toward electrification, push
buttons became desirable at a moment of speculation about
labor-saving inventions and to advance the cause of "domestic
engineering" to reduce housework. As the structure of household
labor began to change, with paid and unpaid servants leaving

for employment outside of the home, housewives faced an increase in daily tasks and confronted a new set of tools to manage these tasks.[7] Many of these technologies, however, often made more work for the women whom they were invented to spare.[8] Electricians and manufacturers argued for button-activated devices as part of technologically advanced homes that predated how we think about "smart homes" today, and plans for the electrified household fed off fantasies of domestic control and effortlessness.[9]

Given that pushing buttons reflected a common communication practice, calling or signaling to get attention or to achieve a particular effect, it is essential to consider push buttons' place in a broader revolution in communication technologies in the late nineteenth century. Historians have extensively documented two major innovations, telegraphy and telephony, at this historical moment.[10] Discussing how society incorporated, repurposed, and imagined telegraphs and telephones for business and personal use, such studies reveal changing patterns in communication across distance as users grappled with sending and receiving messages in non-face-to-face contexts. Internal communication systems in offices also underwent a transformation based on a new philosophy of management that incorporated typewriters, filing systems, and archival methods quite different from informal strategies of the past.[11] Importantly, however, push buttons played a role somewhere among telegraphs, telephones, and internal communication systems, making the concept of "calling" for people and things over short distances and within closed-circuit spaces both prevalent and familiar. As popular one-way communication devices, buttons often garnered significant attention alongside these other tools, although they have received scant mention from those chronicling the past. Nevertheless, according to one electrician (1909) writing on these technologies, "They have all become electrical matters of course like the telegraph, the telephone, and the ubiquitous push button."[12] Later, as telephones gained traction in households, promoters encouraged people to switch from push buttons to telephones for their communication needs or to internal early intercom systems (originally called "intercommunicating systems") that featured push buttons. This shift suggests that users' familiarity with buttons actually helped to pave the way for increased uptake of telephones, and thus push buttons played a nontrivial role in telephone consumption.

Although much room remains to incorporate push buttons and their associated hand practices into these broader historical narratives, it is not the case that push buttons have gone totally unnoticed. Influential thinkers in fields such as media studies and philosophy of technology have considered the implications of a society so intoxicated with button pushing to some extent. These works, however, yield a uniform perspective that describes pushing a button as a lack of authentic engagement with the world. Baudrillard (1968) suggests that buttons, levers, and the like have made it so that button pushers are "actors in a global process in which *man* is merely the role, or the spectator" (italics original).[13] Meanwhile, De Certeau et al. (1998) offer a similarly bleak picture of the person who spends her day pushing buttons: "[She] unleashes the movement by pushing on a button, and collects the transformed matter without having controlled the intervening steps."[14] Further, she has "become the *unskilled spectator* who watches the machine function in her place" (italics original).[15] Winner (1978) also concludes that human beings accustomed to increasingly automated technologies have become "in a true sense a redundancy," and their "governance is ornamental rather than decisive."[16] More recently, philosopher Flusser (1999) makes an all-too-common proposition that the "hands have become redundant and can atrophy," leaving behind only fingertips that carry out a "programmed freedom" devoid of true agency.[17] In each of these cases, the act of pushing buttons is positioned in antithesis to legitimate, interactive experiences with technology. By arguing for button pushing as an essentially passive activity, the authors leave little room for the incredible diversity of user experiences at different historical moments, for the ways that push buttons get configured (and in fact not all buttons are alike), and for the position that every activity, every touch a hand undertakes carries with it some significance. This book has engaged such a dialogue by taking seriously how production, consumption, technologies, and meaning come together to produce buttons and button pushers, both past and present.[18]

Notes

Introduction

1. Geo. L. Cooper, "If Ben Had Pressed the Button," *American Economist*, August 25, 1893: 100.

2. Karthik Srinivasan, "The Aftermath of the Buttonization of Our Emotions," *social@Ogilvy*, March 1, 2016, https://social.ogilvy.com/the-aftermath-of-the-buttonization-of-our-emotions/.

3. Janet Zandy, *Hands: Physical Labor, Class, and Cultural Work* (New Brunswick, NJ: Rutgers University Press, 2004).

4. Notable exceptions include: Anne Cranny-Francis, *Technology and Touch: The Biopolitics of Emerging Technologies* (New York: Palgrave Macmillan, 2013); Mark Paterson, *The Senses of Touch: Haptics, Affects and Technologies* (Oxford: Berg, 2007); David P. Parisi, "Tactile Modernity: On the Rationalization of Touch in the Nineteenth Century," in *Media, Technology, and Literature in the Nineteenth Century*, ed. Colette Colligan and Margaret Linley (Farnham, UK: Ashgate, 2011), 189–214; and Colette Colligan and Margaret Linley, *Media, Technology, and Literature in the Nineteenth Century: Image, Sound, Touch* (Farnham, UK: Ashgate, 2011). The latter features a number of chapters on historical narratives of touch and technology, but these essays focus on keys and typing rather than the broader set of changes in hand practices, hand environments, and hand relationships as detailed in *Power Button*.

5. See Carolyn Marvin, *When Old Technologies Were New: Thinking About Electric Communication in the Late Nineteenth Century* (New York: Oxford University Press, 1988).

6. For a useful explication of the term "digital" in the digital age as referring to hands and digital computing, see Benjamin Peters, "Digital," in *Digital Keywords: A Vocabulary of Information Society and Culture*, ed. Benjamin Peters (Princeton, NJ: Princeton University Press, 2016), 93–108.

7. "Command (n)," in *Oxford English Dictionary* (New York: Oxford University Press, 2017).

8. Hensleigh Wedgwood, *Dictionary of English Etymology* (New York: Macmillan, 1878).

9. Charles Dickens, "What There Is in a Button," in *Household Words*, vol. 5, ed. Charles Dickens (London: 1852), 109. See also Nina Edwards, *On the Button: The Significance of an Ordinary Item* (London: I. B. Tauris, 2011), for a historical account of clothing buttons and an extended discussion of their significance.

10. "Art XXX—Scientific Intelligence," *Edinburgh Philosophical Journal* 7 (1822): 385. See also James W. Benson, *Time and Time-Tellers* (London: Robert Hardwicke, 1875).

11. "A New Computer," *Museum of Foreign Literature, Science and Art* 2, no. 10 (1823): 384.

12. "Golden Telegraph Key for President's Use," *New York Times*, May 15, 1927: XX16.

13. Bennet Woodcroft, *Chronological and Descriptive Index of Patents Applied for and Patents Granted* (London: George Edward Eyre and William Spottiswoode, 1874), 145.

14. For example, a brochure advertised that the user could trigger a "simmering fire at will by simply turning the button." See Hughes Electric Heating Co., "The Hughes Electric Cook Stove," 1910 (Warshaw Collection of Business Americana, Archives Center, National Museum of American History, Smithsonian Institution, Washington, DC).

15. One engineer specifically refers to the spinet piano's relationship to push buttons. See "Automatic Devices," *Electrical Age*, 1898: 339. More broadly, we might consider how musical instruments and information and communication technologies relate to one another, particularly in terms of tactile habits. See, for example, Ivan Raykoff, "Piano, Telegraph, Typewriter: Listening to the Language of Touch," in *Media, Technology, and Literature in the Nineteenth Century: Image, Sound, Touch*, ed. Colette Colligan and Margaret Linley (Farnham, UK: Ashgate, 2011), 159–188.

16. In *Populuxe* (New York: Knopf, 1986), Thomas Hine focuses on the Cold War era as representative of the "push-button age."

17. Jussi Parikka, *What Is Media Archaeology?* (Cambridge: Polity Press, 2013), 13. Parikka draws on Foucault's "genealogical approach" to discuss how to conduct historical studies that do not take a typical chronological approach.

18. Westinghouse Electric & Manufacturing Co., "Your Unseen Servant," *Popular Mechanics* 28, no. 1 (1917): 102; "The Unseen Servant behind the Perfect Meal Is the Perfect Refrigerator," *Literary Digest* 61 (1919): 146; Charles P. Steinmetz, "The A-B-C's of Electricity," *Popular Science* 101, no. 1 (1922): 29.

19. This reference to the "inside" of the machine applies to a concept called "black-boxing" in science and technology studies (STS). For one take on this concept as it relates to discourses of magic, see William A. Stahl, "Venerating the Black Box: Magic in Media Discourse on Technology," *Science, Technology & Human Values* 20, no. 2 (1995): 234–258.

20. Stephen Kern, *The Culture of Time and Space, 1880–1918: With a New Preface* (Cambridge, MA: Harvard University Press, 2003), 1. See also Wolfgang Schivelbusch and Damien J. Kulash, *The Railway Journey: Trains and Travel in the 19th Century* (New York: Urizen Books, 1979).

21. Terrell Croft, *Wiring of Finished Buildings: A Practical Treatise Dealing with the Commercial and the Technical Phases of the Subject, for the Central-Station Man, Electrical Contractor and Wireman* (New York: McGraw-Hill, 1915).

Chapter 1: Setting the Stage

1. Ernest Gellner, *Legitimation of Belief* (Cambridge: Cambridge University Press, 1975), 76.

2. See other examples of "mere touch," such as John Milton Gregory, *A New Political Economy* (Cincinnati, OH: Bragg & Co., 1882); "The Wonders of Electricity," *Journal of the Brotherhood of Locomotive Engineers* 24 (1890): 913–914; Alexander Findlay, *Chemistry in the Service of Man* (London, New York: Longmans, Green and Co., 1920).

3. "A Modern Organ," *Journal of the Franklin Institute* 72, no. 1 (1876): 186–191.

4. Francis Trevelyan Miller, *Wonder Stories* (New York: The Christian Herald, 1913).

5. Walter Benjamin, "On Some Motifs in Baudelaire," in *Walter Benjamin: Selected Writings*, ed. Howard Eiland and Michael W. Jennings, vol. 4, *1938–1940* (Cambridge, MA: Harvard University Press), 328.

6. *The Factory Management Series: Machinery and Equipment* (Chicago: A. W. Shaw Company, 1915).

7. Karl Marx and Friedrich Engels, *Manifesto of the Communist Party* (London: Reeves, 1888).

8. Jean Baudrillard, *The System of Objects* (London: Verso, 2005); Michel de Certeau, Luce Giard, and Pierre Mayol, *The Practice of Everyday Life: Living and Cooking*, vol. 2 (Minneapolis: University of Minnesota Press, 1998); Vilém Flusser, *Shape of Things: A Philosophy of Design* (London: Reaktion Books, 1999).

9. Michel Foucault, *Essential Works of Foucault 1954–1984: Power*, vol. 3 (London: Penguin Books, 2000); and Bruno Latour, *Science in Action: How to Follow Scientists and Engineers through Society* (Cambridge, MA: Harvard University Press, 1988).

10. William H. Sewell Jr., "Toward a Post-Materialist Rhetoric for Labor History," in *Labor History: Essays on Discourse and Class Analysis*, ed.

Lenard R. Berlanstein (Champaign-Urbana: University of Illinois Press, 1993), 15–38.

11. Shoshana Zuboff, *In the Age of the Smart Machine: The Future of Work and Power* (New York: Basic Books, 1988).

12. "Work" depends on definition. "What counts" as work might vary significantly across industries and between actors. See Susan Leigh Star and Anselm Strauss, "Layers of Silence, Arenas of Voice: The Ecology of Visible and Invisible Work," *Computer Supported Cooperative Work* 8 (1999): 9–30. For further discussion of the sociology of work, see Grint, "Introduction," in *The Sociology of Work* (Cambridge: Polity Press, 2005), 11. Grint reminds us that "the difference between work and non-work seldom lies within the actual activity itself and more generally inheres in the social context that supports the activity."

13. See Janet Zandy, *Hands: Physical Labor, Class, and Cultural Work* (New Brunswick, NJ: Rutgers University Press, 2004).

14. W. H. Preece, "Recent Wonders of Electricity," *Popular Science* 20, no. 46 (1882): 786; George Walter Stewart, "A Contribution of Modern Physics to Religious Thought," *Homiletic Review* 68 (1914): 278.

15. James W. Steele, *Steam, Steel and Electricity* (Chicago: The Werner Company, 1895).

16. Linda Simon, *Dark Light: Electricity and Anxiety from the Telegraph to the X-Ray* (Orlando, FL: Harcourt, 2004). See also Graeme Gooday, *Domesticating Electricity: Technology, Uncertainty and Gender, 1880–1914* (London: Pickering & Chatto, 2008).

17. "Contagion (n)," in *A Dictionary of the Derivations of the English Language* (London: William Collins, Sons, & Co., 1872), 76.

18. Solomon Solis-Cohen, *A System of Physiologic Therapeutics: Prophylaxis, Personal Hygiene, Civic Hygiene, and Care of the Sick*, vol. 5 (Philadelphia: P. Blakiston's Son & Co., 1903), 289.

19. Augustus Bozzi Granville, *A Letter to the Right Hon. F. Robinson, President of the Board of Trade, and Treasurer of the Navy, on the Plague and*

Contagion with Reference to the Quarantine Laws (London: Burgess and Hill, 1819), 26.

20. J. N. Ritter von Nussbaum, "The Influence of Antiseptics Upon Legal Medicine: A Lecture Delivered in the Course of Clinical Surgery in the Munich Hospital, in the Winter Session 1897–80," *Edinburgh Medical Journal* 26, no. 11 (1881): 1000.

21. Albert John Nunnamaker and Charles Otto Dhonau, *Hygiene and Sanitary Science: A Practical Guide for Embalmers and Sanitarians* (Cincinnati, OH: Embalming Book Company, 1913), 52.

22. "The Theory of Knowledge," *Methodist Review* 46 (1897): 276.

23. Albert Salisbury, *The Theory of Teaching and Elementary Psychology* (Whitewater, WI: The Century Book Company, 1905). See also Grace D. Coleman, "The Efficiency of Touch and Smell," *American Annals of the Deaf* 67 (1922): 301.

24. Andreas Bernard, *Lifted: A Cultural History of the Elevator* (New York: New York University Press, 2014).

25. "The Touch of a Button," *Central Station* 21 (1922):219.

26. John Allen and Chris Hamnett, *A Shrinking World? Global Unevenness and Inequality* (Oxford: Open University and Oxford University Press, 1995), 9. See also other scholars writing on the subject of "time-space compression" at this time period, such as Stephen Kern, *The Culture of Time and Space, 1880–1918: With a New Preface* (Cambridge, MA: Harvard University Press, 2003); Anthony Giddens, *A Contemporary Critique of Historical Materialism: Power, Property and the State* (Berkeley: University of California Press, 1985), 90–108.

27. Bellamy, *Equality* (New York: D. Appleton and Company, 1897), 348.

28. H. P. Bowditch and Wm. F. Southard, "A Comparison of Sight and Touch," *Journal of Physiology* 3 (1882): 233.

29. John Durham Peters, *Speaking into the Air: A History of the Idea of Communication* (Chicago: University of Chicago Press, 1999), 269.

30. George T. Lemmon, *The Eternal Building or the Making of Manhood* (New York: Eaton & Mains, 1899), 158.

31. R. Mullineux Walmsley, *The Electric Current: How Produced and How Used* (London: Cassell and Company, Limited, 1894). By 1917, remote control kinds of switches were extensively used in places like theaters, factories, and central stations. See George A. Schneider, "Technical Hints," *Journal of Electricity* 39, no. 11 (1917): 517–518.

32. Kidder, *Architect's and Builder's Pocket-Book*, 1333. Buttons were frequently lauded for their "neat appearance." See O. S. Platt Manufacturing Company, "The 'New England' Push-Button Switch," *Electrical World* 29 (1897): 695.

33. See Schuylkill Electric Const. & Supply Co., "Specifications for Electrical Work for Mr. Simon Krick," 1909 (Warshaw Collection of Business Americana, Archives Center, National Museum of American History, Smithsonian Institution, Washington, DC).

34. See, for example, the Stout-Meadowcroft Company, *Illustrated Catalogue and Price List of the Stout-Meadowcroft Co.* (New York: Stout-Meadowcroft, 1885; collections of the Bakken Museum, Minneapolis, MN), 106; J. Elliott Shaw Company, "Household and Experimental Electrical Supplies and Novelties, Catalog No. 15," 1903 (Warshaw Collection of Business Americana, Archives Center, National Museum of American History, Smithsonian Institution, Washington, DC).

35. John Wright, *The Home Mechanic: A Manual for Industrial Schools and Amateurs* (New York: E. P. Dutton and Company, 1905), 204.

36. Patrick and Carter Co., *Patrick & Carter's Illustrated Catalogue and Price List* (Philadelphia: Patrick and Carter Co., 1882; Warshaw Collection of Business Americana, Archives Center, National Museum of American History, Smithsonian Institution, Washington, DC).

37. Patrick and Carter Co., *Patrick & Carter's Illustrated Catalogue and Price List* (Philadelphia: Patrick and Carter Co., 1884; Warshaw Collection of Business Americana, Archives Center, National Museum of American History, Smithsonian Institution, Washington, DC).

38. Patrick, Carter & Wilkins Co., *Patrick, Carter & Wilkins Co. Catalogue* (Warshaw Collection of Business Americana, Archives Center, National Museum of American History, Smithsonian Institution, Washington, DC).

39. See other catalogs, including Belden Manufacturing Company, *Belden Manufacturing Company Catalogue*, 1909; Novelty Electric Company, *Novelty Electric Company Illustrated Catalogue*; Ohio Electric Works, *Illustrated Catalogue*, all from Warshaw Collection of Business Americana, Archives Center, National Museum of American History, Smithsonian Institution, Washington, DC.

40. Hart-Hegeman push-button switches, known as "Diamond H Switches," had an international reach, with installations that included the Ritz Hotel, Bank of England, Hotel Maurice and Hotel Regins in Paris, and so on. See "Untitled," *Electrical Review* 38, no. 2 (1901), Warshaw Collection of Business Americana, Archives Center, National Museum of American History, Smithsonian Institution, Washington, DC; Sears, Roebuck and Co., *Electrical Goods and Supplies* (Chicago: Sears, Roebuck and Co., ca. 1902; Collections of the Bakken Museum, Minneapolis, MN).

41. For further discussion of "triggers" and questions of human agency and control, see Keith Grint and Steve Woolgar, "Computers, Guns, and Roses: What's Social about Being Shot?" *Science, Technology & Human Values* 17, no. 3 (1992): 366–380; Jason Puskar, "Pistolgraphs: Liberal Technoagency and the Nineteenth-Century Camera Gun," *Nineteenth-Century Contexts* 36, no. 5 (2014): 517–534.

42. "Untitled," *Electrical Record and Buyer's Reference* 2, no. 1 (1907): 3.

Chapter 2: Ringing for Service

1. George Edwinson, "Electric Bells," *Amateur Work, Illustrated* 1 (1883): 521.

2. Edwinson, "Electric Bells," 323.

3. "Push-Button Inventor Dead," *Sun*, August 17, 1907: 5.

4. T. Commerford Martin and Stephen Leidy Coles, eds., *The Story of Electricity*, vol. 1 (New York: The Story of Electricity Company, 1919).

5. "Electric Fire Alarm; Fifty Years Since the Inauguration of the System," *Sun*, May 1, 1902: 5.

6. Andrew Carnegie, *Triumphant Democracy: Or, Fifty Years' March of the Republic* (New York: Charles Scribner's Sons, 1886), 83.

7. "To Prevent Panic in Theatres," *Harper's Weekly*, April 22, 1882: 243. Although these signals could provide real protection from danger, satirists also mocked them by suggesting that soon enough everyone would have an electric button at hand, with a trapdoor beneath each seat, to carry patrons to safety. The observer humorously noted that a gentleman sitting behind a lady with a tall hat might too make use of this technology, surreptitiously using his button to remove the visual impediment. See "Untitled," *Electrical Journal*, 186. Additionally, push buttons were commonly kept behind glass; the user would break the glass to get to the button. See, for example, "Telephone Fire Alarm," *Municipal Journal & Public Works* 33, no. 21 (1912): 782–783.

8. William Paul Gerhard, "The Essential Conditions of Safety in Theatres.—IV," *American Architect and Building News* 45, no. 969 (1894): 25–26; Gerhard, *Theater Fires and Panics* (New York: John Wiley & Sons, 1897).

9. Hugo Diemer, *Factory Organization and Administration* (New York: McGraw-Hill, 1910). Floor buttons were common in places like offices and hospitals, and shops sometimes placed buttons under doormats to call employees' attention to entering customers. See Rankin Kennedy, *The Book of Electrical Installations*, vol. 2 (London: Caxton Publishing, 1902).

10. William L. Allison, *Allison's Webster's Counting-house Dictionary of the English Language and Dictionary of Electricity, Electrical Terms and Apparatus* (New York: William L. Allison, 1886); Kennedy, *Book of Electrical Installations*; Charles Robert Gibson, *Electricity of To-Day: Its Work & Mysteries Described in Non-Technical Language* (London: Seely, Service & Co. Ltd., 1912).

11. "Great Inventor Hardly Known: Stephen Dudley Field, Man of Wonders," *Boston Daily Globe*, April 17, 1910: SM3; "Electric Bells," *Maine Farmer*, June 29, 1880: 4; Herbert Laws Webb, "The Future of the Telephone Industry," *Engineering Magazine* (1892): 753–761.

12. Charles G. Armstrong, "Improvements in Annunciator and Bell Work," *Electrical Engineer* 12 (1891): 685.

13. "The Electric Call Bells," *Atlanta Constitution*, May 1, 1885: 2.

14. Edwin J. Houston, *A Dictionary of Electrical Worlds, Terms and Phrases* (New York: W. J. Johnston, 1892). See also W. H. McCormick, *Electricity* (London: Frederick A. Stokes, 1915); Charles Tripler Child, *The How and Why of Electricity: A Book of Information for Non-Technical Readers* (New York: D. Van Nostrand Company, 1905); Gibson, *Electricity of To-Day*.

15. G. W. Tunzelmann, *Electricity in Modern Life* (New York: P. F. Collier & Son, 1902), 864. See also Edwin J. Houston, *Electricity in Every-Day Life* (New York: P. F. Collier & Son, 1905).

16. Harold Donaldson Eberlein, "The Revival of Bell Pulls," *House Beautiful* 39, no. 2 (1916): 38–39. See also Mary H. Northend, "Reviving the Bell Pull," *House & Garden* 37, no. 2 (1920):44–60.

17. Eliza Leslie, *The House Book, or, a Manual of Domestic Economy: For Town and Country* (Philadelphia: Carey & Hart, 1845), 332.

18. Mrs. Motherly, *The Servants' Behaviour Book: Or Hints on Manners and Dress for Maid Servants in Small Households* (London: Bell and Daldy, 1859), 29.

19. Edwinson, "Electric Bells," 517–521.

20. Edwinson, "Electric Bells."

21. Eliza Leslie, *Miss Leslie's Behavior Book: A Guide and Manual for Ladies as Regards Their Conversation; Manners; Dress; Introductions; Entree to Society; Shopping; Conduct in the Street; at Places of Amusement. In Traveling; at the Table, Either at Home, in Company, or at Hotels; Deportment in Gentlemen's Society; Lips; Complexion; Teeth; Hands, the Hair; Etc., Etc.* (Philadelphia: T. B. Peterson and Brothers, 1839), 109.

22. Edmund Lester Pearson, "New Books and Old," *Weekly Review* 4, no. 97 (1921): 275.

23. "The Decadence of the Bell: Decline and Fall of an Old-Established Noise-Maker," *Bystander: An Illustrated Weekly* 13, no. 162 (1907): 56. See also W. L. Alden, "Life's Little Worries," *Pearson's Magazine* 5 (1898): 558–559.

24. "Decadence of the Bell," 56.

25. "Decadence of the Bell," 56.

26. Flora Haines Loughead, "The House on the Hill: The Doorbell Tells the Story," *Overland Monthly* 15, no. 85 (1890): 68.

27. John Munro, *The Story of Electricity* (New York: D. Appleton and Company, 1905).

28. William Henry Preece, "On the Application of Electricity to Domestic Purposes," *Telegraphic Journal* 1, no. 16 (1864): 181.

29. "Decadence of the Bell," 56.

30. Fred DeWitt Van Amburgh, *The Buck Up Book* (New York: The Silent Partner Co., 1919), 23.

31. Frederick Niven, *The Porcelain Lady* (London: M. Secker, 1913), 279; Reginald A. R. Bennett, "Electrical Bells: How to Make and Use Them," *Boy's Own Annual* 15 (1893): 717.

32. S. J. Gates, "Electricity in the Home," *Wisconsin Engineer* 20, no. 7 (1916): 301. See also "Introduction of Building Specialties," *American Architect and Architecture* 42 (1893):76.

33. John F. Buchanan, *Brassfounders' Alloys: A Practical Handbook Containing Many Useful Tables, Notes and Data, for the Guidance of Manufacturers and Tradesmen* (London: E. & F. N. Spon, Ltd., 1901), 66.

34. W. W. Atkinson, "Push Versus Pull," *Santa Fe Employees' Magazine* 3 (1909): 1017.

35. James Sully, "Pushing," in *Good Words for 1880*, ed. Donald Macleod (London: Isbister and Company, 1880), 379.

36. Charles Wheatstone, working on telegraphic experiments, noted that "it was possible that one party might be asleep at one end of the wire, he had so arranged the working that the first touch should ring the bell at the other end, even if thousands of miles apart." See Jerrold, *Electricians and Their Marvels* (New York: Fleming H. Revell Company, 1893; collections of the Huntington Library, Pasadena, CA), 156.

37. "Electric Bells," in *Library of Universal Knowledge: A Reprint of the Last (1880) Edinburgh and London Edition of* Chambers's Encyclopædia (New York: American Book Exchange, 1880), 323.

38. John Henry Pepper, *The Boy's Playbook of Science* (London: Routledge, 1881), 235; Eduoard Hospitalier and C. J. Wharton, *Domestic Electricity for Amateurs* (London: E. & F. N. Spon, 1889), 28.

39. "The House Telephone," *House & Garden* (1911): 269. Another source notes that around 1904, larger and newer homes, railroads, and mines were starting to be outfitted with telephones instead of bells and speaking tubes. See American Institute of Electrical Engineers, *The Boston Electrical Handbook: Being a Guide for Visitors from Abroad Attending the International Electrical Congress, St. Louis, Mo.* (Boston: American Institute of Electrical Engineers, 1904; collections of the Bakken Museum, Minneapolis, MN).

40. Christine Frederick, *The New Housekeeping: Efficiency Studies in Home Management* (Garden City, NJ: Doubleday, Page & Company, 1913).

41. David E. Nye, *Electrifying America: Social Meanings of a New Technology, 1880–1940* (Cambridge, MA: MIT Press, 1990); Thomas Parke Hughes, *Networks of Power: Electrification in Western Society, 1880–1930* (Baltimore: Johns Hopkins University Press, 1983).

42. Jacob August Riis, *How the Other Half Lives: Studies among the Tenements of New York* (New York: Charles Scribner's Sons, 1890), 120.

43. Jennie Darlington, "Science for Children," *Pennsylvania School Journal* 38 (1889):170.

44. Charles Barnard, "Some Queer Houses," *Youth's Companion* 65, no. 51 (1892): 2.

45. Barnard, "Some Queer Houses," 2.

46. Barnard, 2. See also McGraw-Hill Book Company, *Wiring Diagrams of Electrical Apparatus and Installations* (New York: McGraw-Hill, 1913).

47. Michael D. Sohon, "Chemistry in Secondary Schools," *Science* 31, no. 808 (1910): 980.

48. Sohon, "Chemistry in Secondary Schools," 980.

49. Isabella Mary Beeton, *Beeton's Housekeeper's Guide; Comprising Complete and Practical Instructions on House Building, Buying, and Furnishing; the Decoration of the Home; the Economical Management of the Household; and the Treatment of Children in Health and Sickness* (London: Ward, Lock, and Co., ca. 1890), 128.

50. G. F. Steele, *Electricity in a Modern Residence* (New York: H. Ward Leonard & Co., 1892). In a treatise on electric bells, Hasluck noted that, "it does not require much ingenuity to construct something that will act as well as the usual form of push." See Paul N. Hasluck, *Electric Bells: How to Make and Fit Them; Including Batteries, Indicators, Pushes, and Switches* (Philadelphia: David McKay, 1914), 128.

51. Christine Terhune Herrick, *The Expert Maid-Servant* (New York: Harper & Brothers, 1904; collections of the Winterthur Library, Wilmington, DE).

52. T. M. Clark, *The Care of a House; a Volume of Suggestions to Householders, Housekeepers, Landlords, Tenants, Trustees, and Others, for the Economical and Efficient Care of Dwelling-Houses* (New York: Macmillan, 1903).

53. Edward Spon and Francis N. Spon, *Spons' Mechanics' Own Book: A Manual for Handicraftsmen and Amateurs* (London: E. & F. N. Spon, 1886), 634.

54. Archibald Williams, *How It Works; Dealing in Simple Language with Steam, Electricity, Light, Heat, Sound, Hydraulics, Optics, Etc. And with Their Application to Apparatus in Common Use* (New York: Thomas Nelson & Sons, ca. 1910), 120.

55. "New Method of Hanging Bell-Wires," *Magazine of Domestic Economy* 6 (1840): 123–124. See also John Wright, "Bell-Hanging for Inside Rooms," *Building Age* 6, no. 2 (1884): 28.

56. The Stout-Meadowcroft Company, *Illustrated Catalogue and Price List of the Stout-Meadowcroft Co.* (New York: Stout-Meadowcroft, 1885), 94.

57. Stout-Meadowcroft, *Illustrated Catalogue and Price List*, 94.

58. John Wright, *The Home Mechanic: A Manual for Industrial Schools and Amateurs* (New York: E. P. Dutton and Company, 1905), 204.

59. See, for example: Stanley Anthony, "Push Button," Patent No. 1,340,139, May 18, 1920; and Cyprien Mailloux, "Push Button," Patent No. 575,523, January 19, 1897.

60. William Henry Preece, "On the Application of Electricity to Domestic Purposes," *Telegraphic Journal* 1, no. 16 (1864): 181; "O'Brien Push-Button Switch," *Motor* 15, no. 5 (1911): 125.

61. T. C. Martin, "The Work and Responsibilities of the 'Local Electrician,'" *Electrical Engineer* 10, no. 133 (1890): 568.

62. See Armstrong, "Improvements in Annunciator," 685; V. A. Kuehn, *Bells and Annunciators Architect and Engineer* 44, no. 1 (1916): 117.

63. William Maver Jr., *William Maver's Wireless Telegraphy: Theory and Practice* (New York: Maver Publishing Company, 1904), 15.

64. William A. Wittbecker, "Domestic Electrical Work," *Sanitary and Heating Age (The Metal Worker)* 43, no. 14 (1895): 86.

65. Mark Peters, *Japan Dreams: Notes from an Unreal Country* (Bloomington, IN: Booktango, 2013), 146.

66. George Heli Guy, "Electricity in the Household," *Chatauquan* 26, no. 1 (1897): 50.

67. Electric Construction and Supply Co., "Private Residences and Apartments," n.d. (Warshaw Collection of Business Americana, Archives Center, National Museum of American History, Smithsonian Institution, Washington, DC).

68. An example of an inexpensive alarm can be found here: Moore's Burglar Alarm Manufacturing Company, "Moore's Burglar Alarm Manufacturing Co," n.d. (Warshaw Collection of Business Americana, Archives Center, National Museum of American History, Smithsonian Institution, Washington, DC); an expensive one here: Electric Construction and Supply Co., "Private Residences and Apartments."

69. "Clocks Which Furnish Light," in *Our Wonderful Progress: The World's Triumphant Knowledge and Works*, ed. Trumbull White (Springfield, MA: Hampden Publishing Company, 1902), 304.

70. As one domestic handbook noted, burglar bells would startle intruders in ways that they wouldn't startle their owners. See Isabella Mary Beeton, *Beeton's Housekeeper's Guide; Comprising Complete and Practical Instructions on House Building, Buying, and Furnishing; the Decoration of the Home; the Economical Management of the Household; and the Treatment of Children in Health and Sickness* (London: Ward, Lock, and Co., ca. 1890).

71. Stout-Meadowcroft Co., *Illustrated Catalogue and Price List*.

72. Western Electric Manufacturing Co., *Electric Bells and Annunciators for Hotels*.

73. "Electric Safety Cabinet," *Los Angeles Times*, May 31, 1897: 9; "The Cashier's Electric Safety Cabinet," *Electrical Engineer* 23, no. 470 (1897): 488.

74. "To Defeat the Cranks," *Electricity Journal* 2, no. 4 (1896): 552.

75. "To Defeat the Cranks," 2.

76. "Electric Safety Cabinet," 9.

77. "Electric Safety Cabinet," 9. See also Kennedy, *Book of Electrical Installations*.

78. Clark, *Care of a House*. See also a fictional example in which the main character, a child experimenter, implements a burglar alarm: L. Frank Baum, *The Master Key: An Electrical Fairy Tale* (Indianapolis, IN: Bowen-Merrill Company, 1901).

79. M. B. Sleeper, *Electric Bells: A Handbook to Guide the Practical Worker in Installing, Operating, and Testing Bell Circuits, Burglar Alarms, Thermostats and Other Apparatus Used with Electric Bells* (New York: Norman W. Henley, 1917), 5.

80. "Patent Notes," *Popular Mechanics* 8, no. 3 (1906): 372.

81. Edward S. Holden, *Real Things in Nature: A Reading Book of Science for American Boys and Girls* (New York: Macmillan, 1903).

82. Steven M. Gelber, *Hobbies: Leisure and the Culture of Work in America* (New York: Columbia University Press, 1999), 155.

83. As one domestic life manual instructed, "Idleness has no place in the model home, and teach your children to work." See Mary Elizabeth Wilson Sherwood, *Amenities of Home* (New York: D. Appleton and Company, 1881), 128.

84. Gelber, *Hobbies*, 155.

85. Edmund Ironside Bax, *Popular Electric Lighting: Being Practical Hints to Present and Intending Users of Electric Energy for Illuminating Purposes With a Chapter on Electric Motors* (London: Biggs & Co., 1891); Hospitalier and Wharton, *Domestic Electricity for Amateurs*; Edward Trevert, *How to Make and Use an Electric Bell* (Lynn, MA: Bubier Publishing Co., 1906).

86. Guy, "Electricity in the Household," 50; E. S. Greeley, "Electricity Applied to Household Affairs," *Independent: A Weekly Journal of Free Opinion* 45 (1893): 7–8; Iles, "Electricity as a Domestic," *Everybody's Magazine* (1901): 344. For a broader analysis of do-it-yourself culture during this time period, see Steven M. Gelber, "Do-It-Yourself: Constructing, Repairing and Maintaining Domestic Masculinity," *American Quarterly* 49, no. 1 (1997): 66–112. Gelber suggests that housework could be made "acceptable" for men when couched in a do-it-yourself, "Mr. Fixit" culture.

87. Helena Higginbotham, "The Electric Bell a Woman's Charge," *Good Housekeeping* 40 (1905): 642–644.

88. For a discussion of the relationship between anxiety and electricity, see Linda Simon, *Dark Light: Electricity and Anxiety from the Telegraph to the X-Ray* (Orlando, FL: Harcourt, 2004). Simon notes that historical

actors associated electricity with visions of wonder, magic, and haunting, among other things.

89. Simon, *Dark Light*.

90. Ernest B. Kent, "The Elementary School and Industrial Occupations," *Elementary School Teacher* 9, no. 4 (1908): 178–185. Outside of school, Boy Scout organizations also educated male youth about how to construct mechanisms like push buttons. See C. Arthur Pearson, *Things All Scouts Should Know: A Collection of 313 Illustrated Paragraphs of Useful Information, Specially Selected for the Use of Boy Scouts* (London: Limited, 1910).

91. See Hazel W. Severy, "Applied Science as the Basis of the Girl's Education," *Journal of Proceedings and Addresses of the National Education Association of the United States* 53 (1915): 1020–1021; P. Crecelius, "Repairing the Electric Bell," in *The Twentieth Yearbook of the National Society for the Study of Education: Part I: Second Report of the Society's Committee on New Materials of Instruction,* ed. G. M. Whipple (Bloomington, IN: Public School Publishing Company, 1921), 163–164.

92. Hugo Newman, "Science Teaching in Elementary Schools," *Elementary School Teacher* 6, no. 4 (1905): 192–202.

93. James William Norman, *Contributions to Education* (New York: Columbia University Teachers College, 1922).

94. Harry Orrin Gillette, "A Point of View in the Teaching of Electricity in the University Elementary School," *Elementary School Journal* 6 (1906): 306–309.

95. "The Month's Review: What Educational People Are Doing and Saying," *American Educational Review* 31, no. 1 (1909): 3–15.

96. Otis W. Caldwell, "Natural History in the Grades," *Elementary School Teacher* 11, no. 2 (1910): 49–62. See also L. Dow McNeff, "Electricity as a Subject for Study in Elementary Schools. Part I," *Elementary School Teacher* 8, no. 5 (1908): 271–276.

97. Joseph Henry Adams and Joseph B. Baker, *Harper's Electricity Book for Boys* (New York: Harper & Brothers, 1907); Holden, *Real Things in Nature*.

98. M. Thomas St. John, *Things a Boy Should Know about Electricity* (New York: Hard Press, 1900).

99. "A Boy and a Bell," *Atlanta Constitution*, November 4, 1900: B2.

100. J. L. Dickson, "How to Make an Aluminum Push-Button," *Science and Industry* 7 (1907): 42.

101. Adams and Baker, *Harper's Electricity Book*.

102. "Walnuts and Wine," *McBride's Magazine*, 1907: 885.

103. "Rats Derange Electric Bells," *New York Times*, January 21, 1898: 12.

104. "What One Electric Bell Did," *Galveston Daily News*, November 4, 1889: 6.

105. Barnard, "Some Queer Houses," 2.

106. "Odd Phases of Chicago Life as Presented to an Observer in a Single Day," *Chicago Daily Tribune*, May 4, 1896: 1.

107. Conflicts often occurred between expert and layperson groups, creating "in" and "out" groups. See Carolyn Marvin, *When Old Technologies Were New: Thinking About Electric Communication in the Late Nineteenth Century* (New York: Oxford University Press, 1988).

108. "Peculiar Electrical Accident," *Medical News* 85, no. 23 (1904): 1092.

109. James Lee Humfreville, *Twenty Years among Our Savage Indians: A Record of Personal Experiences, Observations, and Adventures among the Indians of the Wild West* (Hartford, CT: The Hartford Publishing Company, 1897), 592.

110. It was almost considered a "disgrace" for an electrician to work on bells, and he must have "courage" to announce it publicly. See Armstrong, "Improvements in Annunciator," 685.

111. Mary Smith Lockwood, *Historic Homes in Washington: Its Noted Men and Women* (New York: Belford Company, 1889), 147.

112. Hunter, "Push Button."

113. Oehring, "Circuit-Closer"; McLaughlin, "Push Button."

114. Rudolph M. Hunter, "Push Button," Patent No. 510. December 12, 1893: 540.

115. Hasluck, *Electric Bells.*

116. "Combination Closed and Open Car of Pay-as-You-Enter Type for the Third Avenue Railroad Company, New York," *Electric Railway Journal* 33, no. 4 (1909): 137–141.

117. "The Necessity of Lowering the Steps and Push-Buttons of the Local Street Cars," *English Journal* 1 (1912):636–637.

118. "High Bell Buttons," *Popular Mechanics* 10, no. 1 (1908): 132.

119. "Ringing the Bell and Running," 43.

Chapter 3: Servants out of Sight

1. C. A. Martineau, "Royal Victoria Hall," *Knowledge* 11 (1888):137.

2. See Fiona Macdonald, *Victorian Servants, a Very Peculiar History* (Brighton, UK: Salariya, 2011).

3. Susan Leigh Star and Anselm Strauss, "Layers of Silence, Arenas of Voice: The Ecology of Visible and Invisible Work," *Computer Supported Cooperative Work* 8 (1999): 9–30.

4. Macdonald, *Victorian Servants.*

5. "An Improved Push-Button," *Electrical World* 16 (1890): 266.

6. This raises questions about the relationship between the body and electricity, and between the body and technology more broadly. See Carolyn Thomas de la Peña, *The Body Electric: How Strange Machines Built the Modern American* (New York: New York University Press, 2003).

7. In general, however, builders agreed that buttons at front doors should be visible to their users—not hidden behind shadows: "The first

purpose of a push button … is to be useful, and this purpose is best served when it is plainly accessible to those desiring to enter the house." See National Association of Builders, "The Builder's Exchange," *Building Age* 15 (1893): 113.

8. Western Electric Manufacturing Co., "Electric Bells, Annunciators, Burglar Alarms, Electro-Mercurial Fire Alarm and Electric Gas Lighting" 1882 (Collections of the Huntington Library, Pasadena, CA); J. Elliott Shaw Company, Household and Experimental Electrical Supplies and Novelties, Catalog No. 15," 1903 (Warshaw Collection of Business Americana, Archives Center, National Museum of American History, Smithsonian Institution, Washington, DC).

9. A. J. Wilkinson & Co., *Illustrated Catalogue of Electrical Goods and Bell-Hangers Supplies*, 1891 (Warshaw Collection of Business Americana, Archives Center, National Museum of American History, Smithsonian Institution, Washington, DC); Paul N. Hasluck, *Electric Bells: How to Make and Fit Them; Including Batteries, Indicators, Pushes, and Switches* (Philadelphia: David McKay, 1914); J. H. Bunnell & Co., "Illustrated Catalogue and Price List of Telegraphic, Electrical & Telephone Supplies No. 9," 1888 (Warshaw Collection of Business Americana, Archives Center, National Museum of American History, Smithsonian Institution, Washington, DC). Buttons attached to cords were less likely to break because they could move from place to place and would not get crushed under the weight of a table leg.

10. Bent L. Weaver and William R. Miller, "Push-Button System for Desks and the Like," Patent No. 1,003,677, September 19, 1911.

11. These mechanisms were primarily created for popular gasoline vehicles to make them more like electric vehicles.

12. Francis Cruger Moore, *How to Build a Home: Being Suggestions as to Safety from Fire, Safety to Health, Comfort, Convenience, Durability and Economy* (New York: Doubleday & McClure, 1897), 113.

13. "Untitled," *American Stationer* 29 (1891), 483. Tinkerers also created such devices to experiment with electrical effects. L. Frank Baum's son, for example, documented in his autobiography, "I rigged up an annun-

ciator drop in the kitchen. ... As soon as I got out of bed, I pushed a button in my room and the annunciator came down with a sign saying, 'start breakfast,' This was the signal to our cook, and by the time I got down, my breakfast was ready." See L. Frank Baum, "Preface," in *The Master Key: An Electrical Fairy Tale* (Indianapolis, IN: Bowen-Merrill, 1901).

14. Quoted in "Untitled," *Western Electrician*, 1–2 (1887): 18. The article notes that electricians found this piece in the *Sun* amusing because this was a "minor application" of electricity and not at all complicated, but the author applauds the mention because it would "do no harm" to the electrical supplies business.

15. Roger B. Whitman, "Planning for the Wiring of the House," *Country Life* 40 (1921): 63.

16. "Combination Floor Key and Button," *Western Electrician* 3, no. 18 (1888): 229.

17. Mary Pattison, *Principles of Domestic Engineering; or the What, Why and How of a Home* (New York: The Trow Press, 1915); Frederick Winslow Taylor, *The Principles of Scientific Management* (New York: Harper & Brothers, 1913).

18. John Wright, "Bell-Hanging for Inside Rooms," *Building Age* 6, no. 2 (1884): 28. Homeowners with electricity and hotels commonly installed buttons at bedsides for the reclined caller to push for what (or whom) she needed with "ready access to it." See McLaughlin, "Electrical Call Systems for Hotel," Patent No. 335,604, February 9, 1886. For other bedside push-button products, see Manhattan Electrical Supply Co., "The Philosophy and Practice of Morse Telegraph; Also Illustrations, Descriptions and Price List of Something Electrical for Everybody," n.d. (Warshaw Collection of Business Americana, Archives Center, National Museum of American History, Smithsonian Institution, Washington, DC); Sears, Roebuck and Co., *Electrical Goods and Supplies* (Chicago: Sears, Roebuck and Co., ca. 1902; collections of the Bakken Museum, Minneapolis, MN); J. E. H. Gordon, *Decorative Electricity* (London: Sampson Low, Marston, Searle, & Rivington, 1891, collections of the Bakken Museum, Minneapolis, MN).

20. E. G. Crans, "Luxury in Modern Living," *Puritan* 5 (1899): 227.

21. "Editorial Points," *Boston Daily Globe*, September 18, 1898: 26.

22. "Thought He Was on Pullman; Traveler on the Street Car Confuses the Use of Push Buttons," *Boston Daily Globe*, June 21, 1906: SM16.

23. George Bidwell, *Forging His Own Chains: The Wonderful Life-Story of George Bidwell* (Hartford, CT: The Bidwell Publishing Company, 1890), 183.

24. "Mechanical Improvement for Hotel Dining Rooms," *Popular Mechanics* 9, no. 3 (1907): 326.

25. "Untitled," *Harper's Magazine*, 67 (1883): 321.

26. Morris Phillips, *Abroad and at Home: Practical Hints for Tourists* (New York: Brentano's, 1891), 197.

27. "A Call-Bell Wrinkle," *Boston Daily Globe*, June 20, 1890: 2.

28. For another example, see Philippe Gengembre Hubert, "A Letter to the Rising Generation," *Atlantic Monthly* 107 (1911): 147. The author comments, "When one gets light by pushing a button, heat by turning a screw, water by touching a faucet, and food by going down in an elevator, life is so detached from the healthy exercise and discipline which used to accompany the mere process of living."

29. "Editorial Etchings," *Munsey's Magazine* 6, no. 3 (1891): 373–375.

30. Rankin Kennedy, *The Book of Electrical Installations*, vol. 2 (London: Caxton Publishing, 1902).

31. Isabella Mary Beeton, *Beeton's Housekeepers' Guide; Comprising Complete and Practical Instructions on House Building, Buying, and Furnishing; the Decoration of the Home; the Economical Management of the Household; and the Treatment of Children in Health and Sickness* (London: Ward, Lock, and Co., ca. 1890), 128.

32. Greeley, "Electricity Applied to Household Affairs," *Independent: A Weekly Journal of Free Opinion* 45 (1893): 7.

33. "The Decadence of the Bell: Decline and Fall of an Old-Established Noise-Maker," *Bystander: An Illustrated Weekly* 13, no. 162 (1907): 55.

34. "Untitled," *Street Railway Journal* 5 (1889): 57.

35. "Untitled," 57.

36. "Pullman's New Double Decker Street Car," *Chicago Daily Tribune*, November 6, 1897: 7. One successful experiment with push buttons occurred in Kansas City, where "everything seems to work like a charm." See "Push-buttons on Street Cars," *Washington Post*, April 8, 1894: 10.

37. "About Those Buttons: One Man Who Saw How They Didn't Work Very Well," *Boston Daily Globe*, July 23, 1895: 8.

38. "Oppose Push Buttons on Cars," *Chicago Daily Tribune*, July 15, 1900: 14.

39. Stephen P. Rice, *Minding the Machine: Languages of Class in Early Industrial America* (Berkeley: University of California Press, 2004).

40. For further discussion of "backstage" conditions, see Mikhail Bakhtin, *Rabelais and His World*, translated by H. Iswolsky (Bloomington: Indiana University Press, 1993).

41. "Telephones in Industrial Establishments," *Telephony* 9, no. 6 (1905): 508.

42. "He Presses the Button," *Boston Daily Globe*, March 18, 1895: 10.

43. "He Presses the Button," 10.

44. John Durham Peters importantly points out that "one-way communication is not necessarily bad. Reciprocity can be violent as well as fair. ... To say, then, that modes of communication that involve a one-way dispersion are necessarily flawed or domineering is to miss one of the most obvious facts of ethical experience: the majesty in many cases of nonresponsiveness." See Peters, *Speaking into the Air: A History of the Idea of Communication* (Chicago: University of Chicago Press, 1999), 57. In the case of call buttons, however, other facts including physical separation, hierarchical management, bureaucratic procedures, and unequal

hand practices contributed to perceptions of one-way communication as demeaning.

45. "The Miniature Telegraph," in *The Science Record for 1874: A Compendium of Scientific Progress and Discovery During the Past Year*, ed. Alfred E. Beach (New York: Munn & Company, 1874), 111–113.

46. "To Attract Page's Attention," *Boston Daily Globe*, January 10, 1900: 7.

47. "To Attract Page's Attention," 7.

48. "Iowa's New State Capitol Building," *Electrical Review* (1884): 7. Years later, a similar observation stated, "In the Iowa Legislature the good old days have gone out when a loud-voiced clerk called the roll and the members shouted their answers. Now it is a matter of pressing a stealthy little electric button and the thing is done, and from fifteen minutes to a half-hour saved in taking a roll call vote." See "Voting by Button," *Independent* 105, no. 3769 (1921): 461.

49. "Congressmen and Their Electric Lights and Fans," *Electrical Review* 32, no. 19 (1898): 305.

50. "Electrical Aids to Legislation," *Electricity Journal* 1, no. 1 (1895): 55.

51. For further description of tool department use of calling annunciators, see Hugo Diemer, *Factory Organization and Administration* (New York: McGraw-Hill, 1910). See also "Push Button System for Saving Time of Machine Men," *Magazine of Business* 9, no. 6 (1906): 649.

52. "Thinking in Terms of —," *Cartoons Magazine*, 1918: 51.

53. "New Page Call in Operation," *Washington Post*, December 25, 1901: 9.

54. "New Page Call," 9.

55. "The Major Sat on the Push-Buttons," *Sun*, April 11, 1904: 11.

56. "Pushed the Crank Button," *Life* 47, no. 1232 (1906): 708.

57. Chapter 3 discusses this issue of power relations and push-button abuse in depth.

58. Sigfried Giedion notes that domestic mechanization arose in part to combat servant labor (viewed as slave labor by many segments of society). However, the case of the push button contradicts this idea by suggesting that mechanical buttons acted as a further means of enslavement. See Giedion, *Mechanization Takes Command: A Contribution to Anonymous History* (Oxford: Oxford University Press, 1948), 715.

59. Frank Townsend Lent, *Sound Sense in Suburban Architecture: Containing Hints, Suggestions, and Bits of Practical Information for the Building of Inexpensive Country Houses* (New York: W. T. Comstock, 1895), 96.

60. "New System for Hospital Signal Lighting," *Electrical Review and Western Electrician* 67, no. 21 (1915): 934; "An Alarming Alarm for the Burglar: He Could Never Turn It Off," *Popular Science* 91, no. 1 (1917): 49.

61. Janet Floyd, *Domestic Space: Reading the Nineteenth-Century Interior* (Manchester, UK: Manchester University Press, 1999). See also Rudolph M. Hunter, "Push Button," Patent No. 510, December 12, 1893: 540. Others invented similar buttons to deal with continuous ringing. See Louis F. Johnson, "Electric Push-Button," Patent No. 631,892, 1899.

62. Charles Robert Gibson, *Electricity of To-Day: Its Work & Mysteries Described in Non-Technical Language* (London: Seely, Service & Co. Ltd., 1912), 167. In nondomestic environments, such as offices, buzzers and other quieter technologies would be put in place. See W. H. McCormick, *Electricity* (London: Frederick A. Stokes, 1915).

63. See Carroll Westall, *The House Electrical; Being a Brief Description of the Ideal Home and How to Plan and Equip It* (Boston: Pettingell-Andrews, 1912; collections of the Winterthur Library, Wilmington, DE). Regarding problems with hired help at the turn of the twentieth century, see I. M. Rubinow and Daniel Durant, "The Depth and Breadth of the Servant Problem," *McClure's Magazine* 34, no. 5 (1910): 576.

64. G. L. Hoadley, "Home Electrics," *New Science and Invention in Pictures* 9, no. 1 (1921): 430.

65. John Wright, *The Home Mechanic: A Manual for Industrial Schools and Amateurs* (New York: E. P. Dutton and Company, 1905), 204–205.

Chapter 4: Distant Effects

1. James W. Wall, "Address of James W. Wall," in *Appendix to the Journal of the Sixteenth Senate of the State of New Jersey* (Belvidere, NJ: John Simerson, 1860), 562.

2. "Humanisms," *Clothing Trade Journal* 19, no. 2 (1921): 84. The article cites this quote as coming from noted philosopher Voltaire.

3. "Electrical Celebration of New Year's Eve," *Electrician* 14 (1885): 216.

4. William J. Hammer, *Electric Diablerie: Being a Veracious Account of an Electrical Dinner Tendered in 1884 by William J. Hammer, Consulting Electrical Engineer to the "Society of Seventy-Seven" of the N.P.H.S. of Newark, N.J., in the First Electrical House Ever Established Anywhere in the World* (New York: William J. Hammer, 1885).

5. For another example of using a secret electric push button to surprise witnesses with magic, see Albert A. Hopkins, *Magic: Stage Illusions and Scientific Diversions* (London: Sampson Low, Marston, and Company, 1897).

6. F. J. Masten, "Two Speeds for an Engine," *Wood-worker* 12 (1893): 15.

7. William M. Brock, *Electricity in Paterson: Being a Treatise for Every-Day Folk on the Use of Electric Units and the Practical Application of Electric Currents* (Paterson, NJ: Press of the Sunday Chronicle, 1896, collections of the Huntington Library, Pasadena, CA).

8. "The Big Blast," *Los Angeles Times*, October 11, 1885: 1.

9. B. E. Dawson, "My Front Door Key," *Medical Brief* 35 (1907): 901.

10. Elisha Gray, *Nature's Miracles: Familiar Talks on Science*, vol. 2 (New York: Baker & Taylor, 1900), 53.

11. Adolphus Frederick Schauffler, *Select Notes on the International Sunday School Lessons* (Cambridge, MA: W. A. Wilde Company, 1911), 338.

12. Bruno Latour, "Mixing Humans with Non-Humans: Sociology of a Door-Closer," in *Ecologies of Knowledge—Work and Politics in Science and*

Technology, ed. Susan Leigh Star (Albany, NY: SUNY Press, 1995), 257–280. See also Latour, "Force and Reason of Experiment," in *Experimental Inquiries, Historical, Philosophical and Social Studies of Experimentation in Science*, ed. Homer Le Grand (Dordrecht, the Netherlands: Kluwer Academic Publishers, 1990), 48–79.

13. Latour, "Force and Reason of Experiment," 48–79.

14. Jeffrey Sconce, *Haunted Media: Electronic Presence from Telegraphy to Television* (Durham, NC: Duke University Press, 2000), 7.

15. Judith A. Adams, "The Promotion of New Technology through Fun and Spectacle: Electricity at the World's Columbian Exposition," *Journal of American Culture* 18, no. 2 (1995):45–55.

16. David E. Nye, *Electrifying America: Social Meanings of a New Technology, 1880–1940* (Cambridge, MA: MIT Press, 1990).

17. See Tony Bennett, *The Birth of the Museum: History, Theory, Politics* (London: Routledge, 1995); Robert W. Rydell, *All the World's a Fair: Visions of Empire at American International Expositions, 1876–1916* (Chicago: University of Chicago Press, 1987).

18. For more specific analyses of these issues, see John G. Cawelti, "America on Display: The World's Fairs of 1876, 1893, 1933," in *The Age of Industrialism in America; Essays in Social Structure and Cultural Values*, ed. Frederic Cople Jaher (New York: Free Press, 1968), 317–363; Ruth Oldenziel, *Making Technology Masculine: Men, Women, and Modern Machines in America, 1870–1945* (Amsterdam: Amsterdam University Press, 1999); Theda Perdue, *Race and the Atlanta Cotton States Exhibition of 1895* (Athens: University of Georgia Press, 2010).

19. See Rachel Plotnick, "Touch of a Button: Long-Distance Transmission, Communication and Control at World's Fairs," *Critical Studies in Media Communication* 30, no. 1 (2012): 52–68.

20. "To Press the Button: President Cleveland, at Gray Gables, Will Start Our Exposition," *Atlanta Constitution*, July 23, 1895: 11; "Opened the Fair," *Washington Post*, June 2, 1905: 1; N. C. Schaeffer, "Grand Opening Ceremony," *Pennsylvania School Journal* 42 (1893): 13.

21. Arthur E. Kennelly, "The Evolution of Electric and Magnetic Physics," in *Evolution in Science, Philosophy, and Art*, ed. Brooklyn Ethical Association (New York: D. Appleton and Company, 1891), 153.

22. Carolyn Marvin, *When Old Technologies Were New: Thinking About Electric Communication in the Late Nineteenth Century* (New York: Oxford University Press, 1988), 151.

23. F. Herbert Stead, "An Englishman's Impressions at the Fair," *Review of Reviews* 8 (1893): 32.

24. Schaeffer, "Grand Opening Ceremony," 13.

25. "Mrs. Cleveland's Touch," *North American*, August 24, 1886.

26. "Widow of President Polk: She Touched the Electric Button and Started the Cincinnati Exposition," *Atchison Daily Globe*, July 21, 1888.

27. Patricia A. Johnston, *Real Fantasies: Edward Steichen's Advertising Photography* (Berkeley: University of California Press, 1997).

28. Graeme Gooday, *Domesticating Electricity: Technology, Uncertainty and Gender, 1880–1914* (London: Pickering & Chatto, 2008); Oldenziel, *Making Technology Masculine*; Thomas J. Schlereth, *Victorian America: Transformations in Everyday Life, 1876–1915* (New York: HarperCollins, 1991).

29. "A Button for the Baby," *Los Angeles Times*, September 18, 1895: 1; "Baby Marion May Press the Button: Little Daughter of President Cleveland to Start Atlanta's Exposition," *Chicago Daily Tribune*, July 23, 1895; "Her Dainty Touch," *Los Angeles Times*, July 24, 1895: 1.

30. Yone Nogochi, *The American Diary of a Japanese Girl* (New York: Frank A. Stokes, 1902), 46.

31. "May He Touch the Right Button," *Semi-Weekly Tribune*, May 11, 1897: 4.

32. "May He Touch," 4.

33. "The Pan-American Fair: Mr. McKinley and the Presidents of Other Republics Will Start the Machinery on May 1," *New York Times*, March 15, 1901: 9.

34. "Fair Opened in a Blaze of Glory," *Boston Daily Globe*, May 1, 1904: A7.

35. "A Practical Joker," *Sentinel*, May 15, 1887: 4.

36. "Successor of the Chestnut Bell," *Evening Bulletin* (1886): 4.

37. Charles S. Day, *A New Illustrated and Descriptive Catalogue, Illustrating and Describing Many Attractive Novelties, Electrical Goods, Useful Articles, Fancy Goods, Fine Jewelry, &C.* (New Market, NJ: Charles S. Day, ca. 1890; collections of the Winterthur Library, Wilmington, DE).

38. See, for example, C. H. W. Bates & Co., "Electric Chestnut Bell," in *C.H.W. Bates & Co. Big Bargain Catalog* (Boston: C. H. W. Bates & Co., n.d.; Warshaw Collection of Business Americana, Archives Center, National Museum of American History, Smithsonian Institution, Washington, DC).

39. "The Electric Button," *Puck* 21, no. 521 (1887): 219.

40. The Stout-Meadowcroft Company, *Illustrated Catalogue and Price List of the Stout-Meadowcroft Co.* (New York: Stout-Meadowcroft, 1885),

41. Ohio Electric Works, "Electric Light for the Necktie," 1897 (Warshaw Collection of Business Americana, Archives Center, National Museum of American History, Smithsonian Institution, Washington, DC).

42. Ohio Electric Works, "The $3 Electric Light for Necktie or Coat," n.d. (Warshaw Collection of Business Americana, Archives Center, National Museum of American History, Smithsonian Institution, Washington, DC.)

43. Reginald A. R. Bennett, *How to Make Electrical Machines* (New York: Frank Tousey, 1902), 42.

44. Bennett, *How to Make Electrical Machines*, 42.

45. Ardee Manufacturing Co., *Ardee Manufacturing Co. Illustrated Catalogue: Manufacturers, Importers and Jobbers of Toys, Novelties and Mail Order Merchandise* (Stamford, CT: Ardee Manufacturing Co., ca. 1903; collections of the Baker Library, Harvard Business School, Boston, MA).

46. John M. Munro, *The Romance of Electricity* (London: Religious Tract Society, 1893).

47. George M. Hopkins, *Home Mechanics for Amateurs* (New York: Munn & Co., 1903).

48. Richard Briggs Stott, *Jolly Fellows: Male Milieus in Nineteenth-Century America* (Baltimore: Johns Hopkins University Press, 2009), 5.

49. George Woodward Warder, *The Cities of the Sun* (New York: G. W. Dillingham, 1901), 63.

50. Rev. T. N. Toller, *Sermons on Various Subjects* (London: B. J. Holdsworth, 1824), 34.

51. "Touching the Button," *Independent*, May 11, 1893: 10.

52. Gerald Stanley Lee, *The Voice of the Machines: An Introduction to the Twentieth Century* (Northampton, MA: Mount Tom Press, 1901; collections of the Huntington Library, Pasadena, CA), 142.

53. Marvin, *When Old Technologies Were New*; Nye, *Electrifying America*; Sconce, *Haunted Media*; Linda Simon, *Dark Light: Electricity and Anxiety from the Telegraph to the X-Ray* (Orlando, FL: Harcourt, 2004).

54. Emma Curtis Hopkins, *Spiritual Law in the Natural World* (Chicago: Purdy Publishing, 1894), 90.

55. H. Beebe, "Physiology: The Organic Nervous System," *Medical Century: An International Journal of Homeopathic Medicine and Surgery* 6, no. 7 (1898): 207–212.

56. Charles Clarence Batchelder, "The Grain of Truth in the Bushel of Christian Science Chaff," *Popular Science* 72, no. 13 (1908): 211–223.

57. George William Winterburn, "Microbes and Disease. Simple Explanations," *Everybody's Magazine* 3 (1900): 457.

58. Andrew Taylor Still, *Osteopathy, Research and Practice* (Kirksville, MO: A. T. Still, 1910).

59. Robert T. Morris, "Is Evolution Trying to Do Away with the Clitoris?" *American Journal of Obstetrics and Diseases of Women and Children* 26 (1892): 847–858.

60. "Clippings and Comments," *Journal of Orificial Surgery* 1, no. 9 (1893): 661.

61. Rachel Maines, *The Technology of Orgasm: "Hysteria," the Vibrator, and Women's Sexual Satisfaction*, Johns Hopkins Studies in the History of Technology, No. 24 (Baltimore: Johns Hopkins University Press, 1999), 9.

62. Benj. E. Dawson, "Circumcision of Females," *Medical Council* 20 (1915): 52.

63. For discussion of some of the ways that man-machine metaphors circulated, see Lissa Roberts, "The Death of the Sensuous Chemist: The 'New' Chemistry and the Transformation of Sensuous Technology," in *Empire of the Senses: The Sensual Culture Reader*, ed. David Howes (Oxford: Berg, 2005), 503–529. For a later citation of Morris' work, see L. A. Stone, ed., *Sex Searchlights and Sane Sex Ethics: An Anthology of Sex Knowledge* (Chicago: Science Publishing Company, 1922).

64. J. F. Sullivan, "The End of War," *Strand Magazine* 3 (1892): 646.

65. "An Electrical Fancy," *Outlook* 54, no. 4 (1896): 174.

66. "Electrical Fancy," 174.

67. Marvin notes that in reaction to this perilous scenario, the collective response was "startling passivity"; in other words, men would "lose all appetite" for warfare if killing could be made as simple as the push of a button. See Marvin, *When Old Technologies Were New*, 148.

68. Julius Robert Mayer, quoted in Andreas Bernard, *Lifted: A Cultural History of the Elevator* (New York: New York University Press, 2014), 172.

69. George Herbert Palmer, *The Nature of Goodness* (Boston: Houghton Mifflin Company, 1903), 108.

70. A. R. Wallace, "Immortality and Morality," *Borderland* 2 (1895): 10.

71. Many of the fears expressed in the late 1800s reappear in the 1940s and 1950s regarding push-button warfare. See, for example, Joseph Alsop and Stewart Alsop, "Are We Ready for Push-Button War?" *Saturday Evening Post* 220, no. 10 (1947): 18–104; John G. Norris, "Calamity Howlers and Pollyannas Both Wrong on 'Push-Button War,'"

Washington Post, August 3, 1947: B11; Cabell Phillips, "Why We're Not Fighting with Push Buttons," *New York Times*, July 16, 1950: SM7. For a contemporary analysis, see Rachel Plotnick, "Predicting Push-Button Warfare: U.S. Print Media and Conflict from a Distance, 1945–2010," *Media Culture & Society* 34, no. 6 (2012): 655–672.

72. Charles Morris, *The Nation's Navy: Our Ships and Their Achievements* (Philadelphia: J. B. Lippincott Company, 1898), 156.

73. Morris, *The Nation's Navy*, 157.

74. Thos. D. Lockwood, "Electrical Killing," *Electrical Engineer* 7, no. 75 (1888): 89–90.

75. Lockwood, "Electrical Killing," 90. At a meeting of electricians to discuss the constitutionality of the death penalty, one electrician like-wise noted that this form of capital punishment was a "degrading use of electricity." See Ferdinand A. Wyman, "Constitutionality of Execution by Electricity," in *Proceedings of the National Electric Light Association* (New York: James Kempster Printing Company, 1890), 156.

76. William A. Anthony quoted in Wyman, "Constitutionality of Execution," 151.

77. Katy Ryan, *Demands of the Dead: Executions, Storytelling, and Activism in the United States* (Iowa City: University of Iowa Press, 2012).

78. "Modern Woman and Her Place in Social System," *Woman's Athenaeum* 9 (1912): 253–254.

79. Ryan, *Demands of the Dead*.

80. "Passing Events," *Belford's Magazine* 1 (1888): 402–408.

81. "Electrical Executions," *Electricity Journal* 1, no. 4 (1895): 73.

82. Das Telephon, "American Notes," *Telegraphic Journal and Electrical Review* 24, no. 582 (1889): 72.

83. Quoted in Mark Essig, *Edison and the Electric Chair: A Story of Light and Death* (New York: Walker, 2003), 292. Others in the electrical indus-

try also opposed the death penalty, in part, because they did not believe experts should have to carry out executions, and they worried that no ordinary sheriff would want to press the button. See Wyman, "Constitutionality of Execution."

84. Essig, *Edison and the Electric Chair*.

85. "Death Shadowing Them," *New York Times*, July 6, 1891: 1.

86. "Wants to Press the Button," *Washington Post*, February 9, 1892: 1.

Chapter 5: We Do the Rest

1. Copp, "Is It Economy," 6–18.

2. "Untitled," *Electrical Record* 2, no. 1 (1909): 3.

3. For further historical discussion of the concept of automaticity, see Lisa Gitelman, *Scripts, Grooves and Writing: Representing Technology in the Edison Era* (Stanford, CA: Stanford University Press, 1999), 190–205.

4. General Electric Company, *Electricity on the Farm*, 1913 (Warshaw Collection of Business Americana, Archives Center, National Museum of American History, Smithsonian Institution, Washington, DC). For an example of a product that promised to remove complication, see General Electric Company, "Christmas Electric Lighting," ca. 1910 (Warshaw Collection of Business Americana, Archives Center, National Museum of American History, Smithsonian Institution, Washington, DC).

5. Otis Elevator Company, "87 Years of … Vertical Transportation with Otis Elevators," 1939 (Warshaw Collection of Business Americana, Archives Center, National Museum of American History, Smithsonian Institution, Washington, DC).

6. "Elevators for a Palace," *Washington Post*, September 8, 1901: 22. By 1901, the *New York Tribune* estimated that 75 push-button elevators existed in New York City. See also Otis Elevator Company, "87 Years," 238.

7. Frank J. Sprague, "Electric Elevators with Detailed Description of Special Types," paper presented at the 102nd meeting of the American Institute of Electrical Engineers, 1896 (Warshaw Collection of Business Americana, Archives Center, National Museum of American History, Smithsonian Institution, Washington, DC), 1.

8. Edward Charles Robert Marks, *Notes on the Construction of Cranes and Lifting Machinery* (Manchester, UK: Technical Publishing Co., 1904).

9. Herzog Telesme, "Untitled," n.d. (Warshaw Collection of Business Americana, Archives Center, National Museum of American History, Smithsonian Institution, Washington, DC).

10. "The Modern City House," *Building Age* 24 (1902): 158.

11. Henry Smith Williams and Edward H. Williams, *Every-Day Science*, vol. 9 (New York: The Goodhue Company, 1910), 172; Henry Smith Williams and Edward Huntington Williams, "The Modern Sky-Scraper," *Wonders of Science in Modern Life* 5 (1912): 123.

12. Carolyn Thomas de la Peña, *The Body Electric: How Strange Machines Built the Modern American* (New York: New York University Press, 2003).

13. "A Child Can Operate the Otis Elevator," *New York Times*, May 29, 1898: IMS16; "Otis Elevators for Use in Residences," *Philistine* 17 (1903): xii; "Untitled," *American Architect* 95, no. 1741 (1909): 37. In one ad, Otis specifically called attention to the fact that pushing an elevator was much like turning on an electric light. See "Otis Residence Elevators," *American Architect* 96, no. 1775 (1909): 5.

14. "Untitled," *American Architect*, 37.

15. Henry Smith Williams and Edward Huntington Williams, *A History of Science: The Conquest of Nature*, vol. 9 (New York: Harper & Brothers, 1910).

16. "Otis Automatic Electric Elevators," *American Architect and Building News* 81, no. 1483 (1903): xvi. Another source, a treatise on the value of electrical appliances, made similar arguments: H. Ward Leonard, *Electricity in a Modern Residence* (New York: H. Ward Leonard & Co., 1892).

17. "His Own Elevator Operator," *Los Angeles Times*, August 21, 1898: B3.

18. For elevator history, see Andreas Bernard, *Lifted: A Cultural History of the Elevator* (New York: New York University Press, 2014).

19. George A. Hool, *Reinforced Concrete Construction* (New York: McGraw-Hill, 1927), 1443.

20. "The Automatic Store: A Modern Mint," ca. 1910 (Warshaw Collection of Business Americana, Archives Center, National Museum of American History, Smithsonian Institution, Washington, DC).

21. Kerry Segrave, *Vending Machines: An American Social History* (Jefferson, NC: McFarland, 2002), 6.

22. See, for example, John F. Kasson, *Amusing the Million: Coney Island at the Turn of the Century* (New York: Hill & Wang, 1978); Kathy Lee Peiss, *Cheap Amusements: Working Women and Leisure in New York City, 1880 to 1920* (Philadelphia: Temple University Press, 1986). Both authors discuss how "mass" culture and entertainment became prevalent—particularly for lower and working classes—during this time period.

23. "Another Button Scheme," *Daily Inter-Ocean*, 1892: 5.

24. "Another Button Scheme," 5.

25. For a more in-depth discussion of world's fairs and the use of push buttons, see Rachel Plotnick, "Touch of a Button: Long-Distance Transmission, Communication and Control at World's Fairs," *Critical Studies in Media Communication* 30, no. 1 (2012): 52–68.

26. Arthur F. Wines, "Coin Controlled Machinery," *Sibley Journal of Engineering* 13 (1899): 219–220.

27. "A Nickel in the Slot," *Washington Post*, April 19, 1891: 15.

28. "A Nickel in the Slot," 15.

29. "The Meredith Cigar-Vending Machine," *Scientific American* 95, no. 2 (1906): 31.

30. "An Automatic Milkman," *Washington Post*, May 3, 1908: M2.

31. "The Jolly Fellow All-Iron Vending Machine," *Voice of the Retail Druggist*, November 1910: 487.

32. Autosales Gum & Chocolate, "One at Every Station," 1912 (Warshaw Collection of Business Americana, Archives Center, National Museum of American History, Smithsonian Institution, Washington, DC).

33. Autosales Gum & Chocolate, "One at Every Station."

34. "Automatic Pill Vending," *Medical News* 52 (1888): 307.

35. "Automatic Pill Vending," 307.

36. George Herbert Palmer, *The Nature of Goodness* (Boston: Houghton Mifflin Company, 1903).

37. A. H. McIntire, "Pushbutton Habit," *Electricity Journal* 14, no. 2 (1917): 41.

38. Bernard, *Lifted*, 175.

39. "Some Famous Advertisements, and How They Originated," *Art in Advertising* 4, no. 2 (1891): 80.

40. H. C. Brown, "At the Home of the Kodak," *Harper's New Monthly Magazine* 83 (1891): 20.

41. "The Age of Buttons," *Chicago Daily Tribune*, June 8, 1900: 6.

42. "Some Famous Advertisements," 80.

43. See Reese V. Jenkins, "Technology and the Market: George Eastman and the Origins of Mass Amateur Photography," *Technology and Culture* 16, no. 1 (1975): 1–9.

44. "Don't Be 'A Button Presser,'" *McClure's Magazine* 8, no. 6 (1897): 40.

45. C. M. Hayes, "Minutes of Convention of Photographers' Association of America, Held July 12–17, 1897, at Celeron-on-Chautauqua,

Chautauqua County, NY," *Anthony's Photographic Bulletin* 28, no. 7 (1897): 234.

46. "You Press the Button," *Fame* 6 (1897): 378.

47. Hayes, "Minutes of Convention," 234–239; F. A. Waugh, "Landscape Photography," *Photo-Miniature* 3, no. 1 (1902): 1–36.

48. "A Plea for Button-Pressers," *Photographic Times* 20, no. 464 (1890): 385–386.

49. "Slot Machines to Replace Department Store Clerks," *Chicago Daily Tribune*, May 15, 1910: 23.

50. American Supply Company, "Electric Button," *Popular Mechanics*, March 1906: 382.

51. Watrous, "Push the Button and We Do the Rest," *Domestic Engineering and the Journal of Mechanical Contracting* 51, no. 13 (1910): 9.

52. Forbes MFG, "Push the Button," *Hopkinsville Kentuckian*, February 18, 1913: 8.

53. Royal Easy Chairs, "Push the Button—and Rest," *McClure's Magazine* 44 (1914): 142.

54. "Miscellaneous," *Western Medical Reporter* 15, no. 3 (1893): 72.

55. M. J. Hill, "The Clitoris," *Homoeopathic Journal of Obstetrics, Gynaecology and Pediatrics* 18 (1896): 582; Havelock Ellis, *Studies in the Psychology of Sex: Erotic Symbolism, the Mechanism of Detumescence, the Psychic State in Pregnancy* (Philadelphia: F. A. Davis Company, 1914), 132.

56. "The Effects of Inventions," *Railroad Trainmen's Journal* 13, no. 152 (1896): 735–738.

57. Herbert Newton Casson, *The History of the Telephone* (Chicago: A. C. McClurg & Co., 1910).

58. Will Price, "What Is the Arts and Crafts Movement?" *International Wood Worker* 14 (1904): 354–356. Similar concerns were expressed that all workers would have left to do would be to push a button or pull a

lever, and that buttons were now the "heaviest" instruments laborers
had to contend with. See Lida Parce, *Economic Determinism: Or, the Economic Interpretation of History* (Chicago: Charles H. Kerr & Company,
1913), 152; Percival Roberts Jr., "Faith and Works," *Yearbook of the
American Iron and Steel Institute* (1912): 117.

59. "Keep Down Your 'Overhead Expense,'" *Conductor and Brakeman* 35
(1918): 37.

60. Charles Henry Vail, *Principles of Scientific Socialism* (Chicago: Charles
H. Kerr & Company, 1899).

61. "A Steady Job for the Machinist," *American Machinist* 19, no. 2
(1896): 52–53.

62. S. N. D. North, *The Man and the Machine: A Plea for Industrial Education; an Address at the Commencement of the Pennsylvania Museum and
School of Industrial Art* (Boston: Press of Rockwell and Churchill, 1896;
collections of the Huntington Library, Pasadena, CA).

63. Charles H. Ham, *Mind and Hand: Manual Training the Chief Factor in
Education* (New York: American Book Company, 1900).

64. Ham, *Mind and Hand.*

65. Price, "The Attitude of Manual Training to the Arts and Crafts,"
Proceedings of the Eastern Manual Training Association (1903): 14–20.

Chapter 6: Let There Be Light

1. R. S. Hale, "Unutilized Comforts of Electricity," *American Gas Light
Journal* 73, no. 14 (1900): 530.

2. A. J. Wilkinson Co., *Illustrated Catalogue of Electrical Goods and Bell-Hangers Supplies*, 1891 (Warshaw Collection of Business Americana,
Archives Center, National Museum of American History, Smithsonian
Institution, Washington, DC); Geo. D. Shepardson, *Electrical Catechism:
An Introductory Treatise on Electricity and Its Uses* (New York: American
Electrician Co., 1901).

3. T. M. Clark, *The Care of a House; a Volume of Suggestions to House-holders, Housekeepers, Landlords, Tenants, Trustees, and Others, for the Economical and Efficient Care of Dwelling-Houses* (New York: Macmillan, 1903).

4. Clark, *The Care of a House*. See also "Proceedings of the Western Gas Association," *Proceedings of the Western Gas Association* 15–18 (1900): 267. Speaking at the association's meeting, one man emphasized how the button could ward off burglars and keep him safe with a touch installed at his bedside.

5. Helen Churchill Candee, "House Building," in *The House and Home: A Practical Book*, ed. Lyman Abbott (New York: Charles Scribner's Sons, 1896), 65–100. In a book targeted at gas companies, manufacturers, engineers, dealers, builders, and inspectors, William Paul Gerhard notes that inhaling leaky gas and dying from asphyxiation was a common problem with this kind of energy. See Gerhard, *Gas-Lighting and Gas-Fitting, Including Specifications and Rules for Gas Piping, Notes on the Advantages of Gas for Cooking and Heating, and Useful Hints to Gas Consumers* (New York: D. Van Nostrand Company, 1894).

6. "'Telephos' Redivivus. Mechanical Difficulties Surmounted," *Gas World* (1908): 105; Amon R. Wells, *A Cyclopedia of Twentieth Century Illustrations: New Picture of Truth from Current Events and Recent Inventions and Discoveries* (New York: Fleming H. Revell, 1918). In the latter text, the author notes that button interfaces for gas lighting often didn't perform as expected, and at times only an "expert can bring a response from push buttons that are dead to others." Because these buttons were often mounted to a wood block and used wood screws, they also proved "unsatisfactory" when the wood got damaged or provided an unreliable action. See John Y. Parke, "Electric Push-Button," Patent No. 727996, May 12, 1903.

7. For one description of the battle over electricity versus gas lighting, see Mark Granovetter and Patrick McGuire, "The Making of an Industry: Electricity in the United States," In *The Law of Markets*, ed. Michael Callon (Oxford: Blackwell, 1998), 147–173.

8. Andrew B. Hargardon and Yellowlees Douglas, "When Innovations Meet Institutions: Edison and the Design of the Electric Light," *Administrative Science Quarterly* 46, no. 3 (2001): 476–501.

9. Frank H. Stewart Electric Company, *Our New Home and Old Times* (Philadelphia: Frank H. Stewart Electric Company, 1913; collections of the Huntington Library, Pasadena, CA). Stewart also noted that at this historical moment, few electricians knew how to carry out wiring for lights, and many more were engaged in the practice of installing doorbells.

10. In a dictionary that described electric switches, for example, one author explained that these interfaces functioned similarly for electric light and electric gaslighting. See Edwin J. Houston, *A Dictionary of Electrical Worlds, Terms and Phrases* (New York: W. J. Johnston, 1892).

11. Sears, Roebuck, and Co., *Electrical Goods and Supplies* (Chicago: Sears, Roebuck and Co., ca. 1902; collections of the Bakken Museum, Minneapolis, MN).

12. Geoffrey C. Bowker and Susan Leigh Star, "How to Infrastructure," in *Handbook of New Media*, ed. Leah A. Lievrouw and Sonia Livingstone (Thousand Oaks, CA: Sage, 2002), 230–245; Hughes, *Networks of Power*.

13. Electrical Specialty Co., *Manufacturers and Owners of Woltmann & Trigg's Celebrated "Standard" Single Push Button Flush Switch* (Denver, CO: The Electrical Specialty Co., 1893; collections of the Winterthur Library, Wilmington, DE).

14. Electrical Specialty Co., *Manufacturers and Owners*.

15. Chris Otter, *The Victorian Eye: A Political History of Light and Vision in Britain, 1800–1910* (Chicago: University of Chicago Press, 2008), 232. See also David Salomons, *Electric Light Installations & the Management of Accumulators* (London: Whitaker & Co., 1890).

16. Robert William MacKenna, *The Adventure of Death* (New York: G. P. Putnam's Sons, 1917), 105–106.

17. "From the Editor's Note Book," *Medical Standard* 28, no. 12 (1905): 684.

18. Robert Hammond, *The Electric Light in Our Homes* (London: Frederick Warne and Co., 1884), 82.

19. "O'Brien Push-Button Switch," *Motor* 15, no. 5 (1911): 44.

20. "'Telephos' Redivivus. Mechanical Difficulties Surmounted," *Gas World* (1908): 105.

21. "'Telephos' Redivivus," 105.

22. Granville E. Palmer, "Good Switches Versus Poor Switches," *Electrical World* 58 (1911): 591.

23. Palmer, "Good Switches," 591.

24. H. Bedford-Jones, "The Problem Answerers," *Popular Electricity and the World's Advance* 6 (1913): 146–151.

25. Bedford-Jones, "The Problem Answerers."

26. One observer, Mrs. Arthur Stallard, suggested that in the case of purchasing a new home, that "one of the most difficult things" was to "stand in the empty rooms and mark exactly the spot where the electric wires must emerge for our own particular fittings. We may take it for granted that the last tenant has left them in the wrong place, as it is very rare to find a house where the lighting is perfectly arranged." See Stallard, *The House as Home* (New York: James Pott & Company, 1913; collections of the Winterthur Library, Wilmington, DE), 72.

27. Dorothy James, "The Building of a House," *Christian Union* 40, no. 26 (1889): 845.

28. Philip Coombs Knapp, *Accidents from the Electric Current: A Contribution to the Study of the Action of Currents of High Potential Upon the Human Organism* (Boston: Damrell & Upham, 1890).

29. Rankin Kennedy, *The Book of Electrical Installations*, vol. 2 (London: Caxton Publishing, 1902); Knapp, *Accidents from the Electric Current*.

30. Helen Churchill Candee, "House Building," in *The House and Home: A Practical Book*, ed. Lyman Abbott (New York: Charles Scribner's Sons, 1896) 65–100.

31. Henrietta C. Peabody, ed., "Homemakers' Questions and Answers: A Ready Reference for Those Who Are Building," in *Remodeling, Furnishing, Decorating or Gardening* (Boston: Atlantic Monthly Press, 1918), 72. For further discussion of the concept of "domestication," see Roger Silverstone and Leslie Haddon, "Design and the Domestication of Information and Communication Technologies: Technical Change and Everyday Life," in *Communication by Design: The Politics of Information and Communication Technologies*, ed. Robin Mansell and Roger Silverstone (Oxford: Oxford University Press, 1996), 44–74; Leslie Haddon, "Roger Silverstone's Legacies: Domestication," *New Media & Society* 9, no. 1 (2007): 25–32.

32. Terrell Croft, *Wiring of Finished Buildings: A Practical Treatise Dealing with the Commercial and the Technical Phases of the Subject, for the Central-Station Man, Electrical Contractor and Wireman* (New York: McGraw-Hill, 1915).

33. Croft, *Wiring of Finished Buildings*.

34. R. Mullineux Walmsley, *The Electric Current: How Produced and How Used* (London: Cassell and Company, Limited, 1894). Nearly all push-button switches were of the "flush" type. See, for example, Sears, Roebuck, and Co., *Electrical Goods and Supplies*; and General Electric, "G.E. Specialties," 1906 (Warshaw Collection of Business Americana, Archives Center, National Museum of American History, Smithsonian Institution, Washington, DC). The electric button at the front door (doorbell) "should also be in harmony with the rest of the house." See "Hardware and the Contractor," *National Builder* 57 (1915): 42–47. Regarding "modern" push-button designs: "Exhibit of the Cutler-Hammer Manufacturing Company Chicago Electrical Show," *Electrical West* 22 (1909): 71.

35. "Advertisement 7—Untitled," *American Architect and Building News* 58 (1897): 1142.

36. "Push-Button Switch," *Mechanical News* 21 (1892): 502. See also "The Cutter Push Button Switches," *Electricity: A Popular Electrical Journal* 4, no. 15 (1893): 208.

37. M. B. Sleeper, *Electric Bells: A Handbook to Guide the Practical Worker in Installing, Operating, and Testing Bell Circuits, Burglar Alarms, Thermostats and Other Apparatus Used with Electric Bells* (New York: Norman W. Henley, 1917); Roger B. Whitman, "Planning for the Wiring of the House," *Country Life* 40 (1921): 63.

38. Bryant Electric Company, *Electrical Wiring Specifications for Residences and Apartment Houses* (Bridgeport, CT: Bryant Electric Company, 1914); "Hardware and the Contractor," 44; Paul N. Hasluck, *Electric Bells: How to Make and Fit Them; Including Batteries, Indicators, Pushes, and Switches* (Philadelphia: David McKay, 1914).

39. Elizabeth Whipple, "Some Electrical Conveniences Women Want Contractors to Provide," *National Builder* 62 (1919): 52.

40. Stallard, *The House as Home*, 75.

41. W. R. Hill, "Door Hardware for the Modern Home," *Building Age* 41, no. 8 (1919): 253; F. W. Cohn, "Electrical Push Button," Patent No. 859,367, July 9, 1907.

42. The disadvantage to the "average" push button was that it was liable to become "broken or damaged." See Elijah H. Stanley, "Electric Push-Button," Patent No. 591,895, 1897.

43. O. S. Platt Manufacturing Company, "The 'New England' Push-Button Switch," *Electrical World* 29 (1897): 695; Stallard, *The House as Home*.

44. General Electric, "G.E. Removable Mechanism Switch," in *Juice: Live Information About Electric Goods* (Boston: Pettingell-Andrews, 1912; collections of the Huntington Library, Pasadena, CA), 7.

45. General Electric, "G.E. Removable Mechanism Switch."

46. Some examples include Oehring, "Circuit-Closer," Patent No. 502,749, August 8, 1893; Sopee, "Electric Push-Button," Patent No. 559,416, filed May 17, 1895, issued May 5, 1896; Anthony, "Push Button," Patent No. 1,340,139, filed March 19, 1919, issued May 18, 1920.

47. Cyprien Mailloux, "Push-Button," Patent No. 575,523, January 19, 1897.

48. William T. Comstock, *Two-Family and Twin Houses* (New York: William T. Comstock, 1908); Archibald Williams, *How It Works; Dealing in Simple Language with Steam, Electricity, Light, Heat, Sound, Hydraulics, Optics, Etc. And with Their Application to Apparatus in Common Use* (New York: Thomas Nelson & Sons, ca. 1910).

49. F. C. Allsop, "Electrical Apparatus, Constructing and Repairing.—XXX," *English Mechanic and World of Science and Art* 60 (1894): 149.

50. Whipple, "Some Electrical Conveniences," 52.

51. J. E. H. Gordon, *Decorative Electricity* (London: Sampson Low, Marston, Searle, & Rivington, 1891; collections of the Bakken Museum, Minneapolis, MN), 131.

52. E. D. Weber, *Practical Wiring for Buildings: For Incandescent Electric Lighting, Electric Gas Lighting, Electric Burglar Alarms, Electric House and Hotel Annunciators, Bells, Etc., Etc.* (New York: The Comenius Press, 1895).

53. Hasluck, *Electric Bells*.

54. "Push Button Locked against Interference by Children," *Popular Mechanics* 31 (1919): 120.

55. Gordon, *Decorative Electricity*.

56. "Bogart's New Burglar Alarm and Gas Lighting Apparatus," *Electrical Engineer* 10, no. 119 (1890): 163–164.

57. Grover Brothers, *Popular Electric Lighting, a Few Practical Hints to Present and Intending Users of ... Electrical Energy for Illuminating Purposes and for Power, and Showing Some of the Latest Methods of ... Heating and Cooking by Electricity* (Newark, NJ: Farrand, 1898); Percy E. Scrutton, *Electricity in Town and Country Houses* (London: Archibald Constable and Co., 1898). J. E. H. Gordon also emphasized that homeowners must impress on servants the importance of turning off lights after using them. Gordon, *Decorative Electricity*, 133.

58. Wolfgang Schivelbusch, *Disenchanted Night: The Industrialization of Light in the Nineteenth Century* (Berkeley: University of California Press, 1988), 7.

59. Schivelbusch, *Disenchanted Night*. Regarding "waste," see also Scrutton, *Electricity in Town and Country Houses*; Arthur F. Guy, *Electric Light and Power* (London: Biggs and Co., 1894).

60. A. J. S., "Turn-out-the-Light," *Rotarian*, November 1924: 51–52.

61. A. J. S., "Turn-out-the-Light."

62. "Electrical Economies in the Home," *Indicator Otis* 4, no. 2 (1911): 10.

63. "'H&H' Automatic Door Switches," *Architectural File* (1920): 1602; Morris G. Miller, "Switch Arm for Use on Basement-Light Circuit," *Popular Mechanics* 23, no. 5 (1915): 764.

64. "'H&H' Automatic Door Switches"; Miller, "Switch Arm." See also Grover Brothers, *Popular Electric Lighting*; Scrutton, *Electricity in Town and Country Houses*; Walmsley, *Electric Current*; Electrical Specialty Co., "Manufacturers and Owners of Woltmann." These authors discuss the benefit of switches for individuals returning home late at night and the advantage of not having to "grope" in the dark.

65. General Electric, *The Home of a Hundred Comforts* (Bridgeport, CT: General Electric, ca. 1920).

66. J. H. Bunnell & Co., *Illustrated Catalogue and Price List of Telegraphic, Electrical & Telephone Supplies No. 9*, 1888 (Warshaw Collection of Business Americana, Archives Center, National Museum of American History, Smithsonian Institution, Washington, DC).

67. Edison Electric Illuminating Company, *Solid Comfort, or the Matchless Man: A Modern Realistic Story in Two Parts* (New York: Edison Electric Illuminating Company, 1903; Warshaw Collection of Business Americana, Archives Center, National Museum of American History, Smithsonian Institution, Washington, DC).

68. Edison Electric Illuminating Company, "The Edison Man," 1906 (Warshaw Collection of Business Americana, Archives Center, National Museum of American History, Smithsonian Institution, Washington, DC).

69. M. S. Seelman Jr., "Electric Light Jingles by Edison Electric Illuminating Co. of Brooklyn," 1908 (Warshaw Collection of Business Americana, Archives Center, National Museum of American History, Smithsonian Institution, Washington, DC).

70. Roscoe Gilmore Scott, "Let Us Go Back," *Edison Monthly* 6, no. 4 (1913): 148 (Warshaw Collection of Business Americana, Archives Center, National Museum of American History, Smithsonian Institution, Washington, DC).

71. In another instance, the Eveready Corporation similarly told its target consumer that, "You instinctively dread the dark. You know that danger and discomfort lie there. ... The pressing of a button releases a brilliant stream of white light that is thrown just where it is wanted." See Eveready Corporation, "Don't Grope in the Dark," 1915 (Warshaw Collection of Business Americana, Archives Center, National Museum of American History, Smithsonian Institution, Washington, DC). Yet another promised "no lamps to be cleaned; no smoke or gas to blacken ceilings and soil hangings." See General Electric, "The Mazda Lamp in the Home," 1910.

72. "Have All Your Machines at Your Finger-Tip," *Bakers Review* 33, no. 2 (1916): 58; "Every Department at Your Finger Tip," *Collier's* 50, no. 8 (1912): 30; Croft, *Wiring of Finished Buildings*.

73. "Vest Pocket Light," 32; Howard Electric Novelty Co., "Reliable Electric Vest Pocket," 273.

74. "'Ever Ready' Vest Pocket."

75. Alexander Findlay, *Chemistry in the Service of Man* (London: Longmans, Green and Co., 1920), 32.

76. See other examples of "mere touch": John Milton Gregory, *A New Political Economy* (Cincinnati, OH: Bragg & Co., 1882); "The Wonders of

Electricity," *Journal of the Brotherhood of Locomotive Engineers* 24 (1890): 913–914; Findlay, *Chemistry in the Service of Man*.

77. Cutler-Hammer Manufacturing Company, "Easy—One Hand," *National Electrical Contractor* 16 (1916): 98.

78. "Distant Control of Small Capacity Power and Lighting Circuits," *Industrial World* 47, no. 14 (1913): 411.

79. Electrical Record, *Special Devices for Electrical Uses* (New York: Gage Publishing, 1917), 24.

80. "An Ideal and Economical New Method of Controlling Electricity," *National Electrical Contractor* 8 (1908): 33.

81. "Method of Controlling Electricity," 33.

82. H. W. Young, "Remote-Control Switches," *Electrical Review* 53, no. 21 (1908): 790.

83. "New Electrically Operated Remote-Control Switch," *Electrical Review and Western Electrician* 69, no. 3 (1916): 130.

84. Wm. Roberts, "Appendix D: Distant Control of Valves," *Proceedings of the American Electric Railway Engineering Association* 11 (1913): 458.

85. See "Remote-Control Switches," 73; Jungmann, "Press the Button," 43; Everett, *Chicago's Awful Theater Horror*, 244.

86. C. H. Claudy, "Wiring for Electric Lighting," *Suburban Life, the Countryside Magazine* 19, no. 3 (1914): 141.

87. Quoted in H. B. Brooks, *Electrical Instruments in England. Special Agents Series—No. 55* (Washington, DC: Government Printing Office, 1912), 35.

88. William Perren Maycock, *Small Switches, Etc., and Their Circuits* (London: S. Rentall & Co., 1911).

89. Lynn W. Meekins, "World Markets for American Manufacturers," *Scientific American* 120, no. 1 (1919): 12.

90. Maycock, *Small Switches*.

91. Ruben Alvin Lundquist, *Electrical Goods in Australia. Special Agents Series—No. 155* (Washington, DC: Government Printing Office, 1918), 54; Lundquist, *Electrical Goods in British India and Ceylon. Special Agent Series—No. 213* (Washington, DC: Government Printing Office, 1922), 96.

92. "Tumbler Switches," *Scientific American* 121 (1919): 53.

93. "Increased Convenience Secured by Lever Switches in Lighting Circuits," *Electrical Record* 30 (1921): 419.

Chapter 7: What's a Button Good For?

1. "Press the Button," *Los Angeles Times*, September 29, 1922: II4.

2. See one example of a faulty encounter between a milkman and a patron: William M. Baum, "Address of Welcome," *Annual Report of the Michigan Dairymen's Association* 23 (1907): 21.

3. "In the Wake of a Push Button," *Magazine of Business* 10, no. 1 (1906): 98–99.

4. "In the Wake," 98–99.

5. Henry M. Hyde, "The Automatic General Manager," *Business, a Magazine for Office, Store, and Factory* 19 (1906): 31–36.

6. "The Disadvantages of Electricity. The Call Button Ego," *Railway Journal* 13, no. 9 (1907): 4–5.

7. "Branch Houses Versus Dealers," *Printers' Ink* 85 (1913): 109.

8. Herman J. Stich, "Push-Button Gents," *Los Angeles Times*, January 11, 1923: II4.

9. "The Bored Woman's Lament," *Washington Post*, February 18, 1906: P4.

10. David Graham Phillips, "The Story of George Helm," *Heart International* 22 (1912): 61–62.

11. "Mackenzie's Patent Electric Bell Indicator," *Electrical Review* 16 (1885): 369.

12. "R. & B. Vibrating Push Button," *American Electrician* 6 (1895): 281. See also Manhattan Electrical Supply Co., *Something Electrical for Everybody, Catalogue Twenty-Six* (New York: Manhattan Electrical Supply Co., ca. 1915), 240.

13. Clarence Edward Clewell, *Handbook of Machine Shop Electricity* (New York: McGraw-Hill, 1916), 112.

14. Clewell, *Machine Shop Electricity*, 112.

15. "In the World of Electricity," *New York Times*, September 10, 1895: 3.

16. Western Electric Company, "Intercommunicating Telephone Systems," 1908 (Warshaw Collection of Business Americana, Archives Center, National Museum of American History, Smithsonian Institution, Washington, DC); "The Girlless Telephone," *Los Angeles Times*, April 12, 1903: D11; "The Houts Automatic Telephone System," *Scientific American* 75, no. 15 (1896): 281.

17. Sears, Roebuck, and Co., *Electrical Goods and Supplies* (Chicago: Sears, Roebuck and Co., ca. 1902; collections of the Bakken Museum, Minneapolis, MN).

18. "Gillette Interior Telephone System," *Electrical World* 23 (1894): 685. Push-button telephones for interior use were in fact in circulation as early as 1887, but individuals did not widely adopt them, in part due to competition from the Bell telephone and in part due to the device's primary inventor, Cornelius Herz, who became embroiled in international scandal. See "The Micro-Telephone Push-Button," *Punch*, 1887; "Dr. Herz's New Telephone: Some Further Account of the Push-Button Device," *New York Times*, May 1, 1887; "Herz's Telephone," *New York Times*, May 21, 1887; and "Facts of Interest," *Frank Leslie's Illustrated Newspaper* 64, no. 1653 (1887): 223.

19. "Gillette Interior Telephone System," 685.

20. Harry B. McMeal, "Push Button Telephones," *Telephony* 3–4 (1902): 329–330.

21. "Western Electric Inter-Phone," *Countryside Magazine and Suburban Life* (1914): 221; McMeal, "Push Button Telephones," 330.

22. "Western-Electric Inter-Phones," *Munsey's Magazine* 43, no. 6 (1910): 118; "Telephone Instruments Masquerading," *Popular Mechanics* 18, no. 4 (1912): 496; "Western Electric Inter-Phone," *Country Life in America* (1913): 101. The same kind of campaign applied to office environments as well: "Western Electric Inter-Phone," 214. Munro also noted that with telephones in the home, a mistress could "give orders to her cook" and a master "can instruct his coachman or his manager from his easy chair." See John M. Munro, *The Romance of Electricity* (London: Religious Tract Society, 1893), 298.

23. "Over the Telephone," *Electrical Magazine and Engineering Monthly* 5 (1906): 52.

24. "In the World of Electricity," 3.

25. "Western-Electric Inter-Phones," 118; "Telephone Instruments Masquerading," 496; "Western Electric Inter-Phone," 101; "Western Electric Inter-Phone," 214; Munro, *Romance of Electricity*, 298.

26. Stromberg-Carlson Telephone Company, "Don't Walk—Push *One* Button *Once*—and Talk," *Textile World* 60 (1922): 1336. See also "Inter-Communicating Telephones and Give Him the Thought of 'Talk, Don't Walk,'" *Merchandising Week* 22, no. 6 (1919): 297.

27. Stromberg-Carlson Telephone Company, "Don't Walk."

28. John Durham Peters, *Speaking into the Air: A History of the Idea of Communication* (Chicago: University of Chicago Press, 1999), 56–57.

29. "Pushed Button in Vain," *Boston Daily Globe*, August 19, 1906: SM2.

30. "Ye Genial Host," *Saturday Evening Post* 194 (1922): 46.

31. William Hannum Grubb Bullard, *Naval Electricians' Text and Handbook* (Annapolis, MD: US Naval Institute, 1904). For more information about fire alarms on navy ships, see "Fire Alarm Systems," *Marine View* 39 (1909): 412.

32. Bullard, *Naval Electricians' Text.*

33. Tracy Walling Burns and Julius Martin, *Electrical Installations of the United States Navy: A Manual of the Latest Approved Material, Including Use,*

Operation, Inspection, Care and Management and Method of Installation on Board Ship (Annapolis, MD: US Naval Institute, 1907), 605.

34. Edwin J. Houston, *Electricity in Every-Day Life* (New York: P. F. Collier & Son, 1905), 13.

35. Rankin Kennedy, *The Book of Electrical Installations*, vol. 2 (London: Caxton Publishing, 1902), 384.

36. Charles Morris, *The Nation's Navy: Our Ships and Their Achievements* (Philadelphia: J. B. Lippincott Company, 1898), 157.

37. "Holtzer-Cabot Electric Company," *Automobile* 22, no. 2 (1910): 126.

38. Bullard, *Naval Electricians' Text*.

39. Electric Signal Clock Co. of Harrisburg and Waynesboro, PA. *Illustrated General Catalogue*, 1891 (Warshaw Collection of Business Americana, Archives Center, National Museum of American History, Smithsonian Institution, Washington, DC).

40. "Fire Alarm Shows Exact Location of Fire," *Popular Mechanics* 19, no. 6 (1913): 780.

41. "The Automatic Fire Alarm Button," *Boston Daily Globe*, November 19, 1899: 34.

42. Frank C. Perkins, "Modern German Police Call and Fire Alarm Systems," *Scientific American* 71, no. S1836 (1911): 148.

43. See, for example, William B. Hopkinson, "Electric Fire-Alarm," Patent No. 769,824, September 13, 1904; Lewis G. Rowand, "Fire-Alarm System," Patent No. 477,068, 1892; George F. Milliken, "Fire-Alarm Box," Patent No. 412,971, 1889.

44. "To Call a Policeman," *Weekly Register-Call*, July 19, 1889.

45. Automobile horns were also sometimes employed as tools for railroad signaling. In one case, 10 horns were wired and mounted to telegraph poles to signal the switching crew rather than doing so by hand.

See "Switching Signals Given with Auto Horns," *Popular Mechanics* 24, no. 2 (1915): 241.

46. "What Happens When a Fire Alarm Box Is 'Pulled,'" *New York Times*, June 2, 1901: SM14; "Automatic Fire Alarm to Be Operated over Telephone Wires," *Scientific American* 92, no. 5 (1905): 102; Charles Robert Gibson, *Electricity of To-Day: Its Work & Mysteries Described in Non-Technical Language* (London: Seely, Service & Co. Ltd., 1912).

47. "New Fire-Alarm System: It Is Designed to Meet Needs of Small Cities," *Sun*, September 17, 1906: 10.

48. Gordon, *Decorative Electricity*, 42.

49. A. F. Collins, "Electricity around the House," *Boys' Life* 6, no. 6 (1916): 32.

50. Western Electric Manufacturing Co., "Electric Bells, Annunciators, Burglar Alarms, Electro-Mercurial Fire Alarm and Electric Gas Lighting," 1882 (Collections of the Huntington Library, Pasadena, CA).

51. "The Spectator," *Outlook* 56 (1897): 732.

52. "The Kitty's Secrets Out," *Los Angeles Times*, June 27, 1901: 9; "M'Coy Wants Money: The Private Detective Claims Cash for Spotting Gamblers," *Atlanta Constitution*, March 14, 1896: 3.

53. "200 Captured in Raid," *Sun*, August 28, 1905: 5.

54. "Push Button in Evidence," *Boston Daily Globe*, February 15, 1895: 12.

55. E. Garfield, "The Tyranny of the Horn—Automobile, That Is," *Essays of an Information Scientist* 6, no. 11 (1983): 216–222.

56. "Signalling Methods Definitely Cared For," *Automobile* 22, no. 2 (1910): 125–128.

57. Manhattan Electrical Supply Co., *Catalog of Automobile Supplies*.

58. Motor Car Equipment Company, *Automobile Accessories, Tools & Hardware Shop Equipment*, 1916 (Warshaw Collection of Business Ameri-

cana, Archives Center, National Museum of American History, Smithsonian Institution, Washington, DC); Imperial Automobile Supply Co., "Catalog."

59. "The Wagner Electric Horn," *Autocar* 20 (1908): 156; Manhattan Electrical Supply Co., *Something Electrical for Everybody, Catalogue Twenty-Six* (New York: Manhattan Electrical Supply Co., ca. 1915). So that motorists did not have to throw away their old rubber bulb horns, they could also purchase push-button attachments: "Electric Auto Horn Company," *Automobile* 22, no. 2 (1910): 126.

60. General Automobile Supply, *Catalogue*, 1909 (Warshaw Collection of Business Americana, Archives Center, National Museum of American History, Smithsonian Institution, Washington, DC), 4.

61. General Automobile Supply, *Catalogue*, 4; "The SirenO—the Mile-Ahead Electric Horn," *Motor* 13 (1910) (Warshaw Collection of Business Americana, Archives Center, National Museum of American History, Smithsonian Institution, Washington, DC); "A Practical Electrical Signal Device for Automobiles and Motor Boats," *Popular Mechanics* 15, no. 5 (1911): 771.

62. "Electric Horns at the Show," *Horseless Age* 27 (1911): 202. This report noted that electric horns were generating "animosity" from the public.

63. "Stow-Away Places in Touring Cars: Novel Devices for Packing Impedimenta on Long Trips," *Town and Country*, November 3, 1906: 24 (Warshaw Collection of Business Americana, Archives Center, National Museum of American History, Smithsonian Institution, Washington, DC), 24; "Benjamin Auto Signals and the Button," *American Chauffeur: An Automobile Digest* 3 (1915): 275.

64. "The Small Boy and the Horn Button," *Scientific American* 108, no. 23 (1913): 509.

65. Henry Walker Young, "Electric Automobile Horn," *Popular Electricity and the World's Advocate* 2 (1909): 746; "Switch for Automobile Horn," *Cycle and Automobile Trade Journal* 23 (1919): 218.

66. "Motor Horn Scares Off Burglars," *Popular Mechanics* 18, no. 4 (1912): 582.

67. A. J. Wilkinson & Co., *Illustrated Catalogue of Electrical Goods and Bell-Hangers Supplies*, 1891 (Warshaw Collection of Business Americana, Archives Center, National Museum of American History, Smithsonian Institution, Washington, DC).

68. "Electric Communicator for Instructing Drivers," *Popular Mechanics* 8, no. 2 (1906): 250.

69. Electric Vehicle Company, "Columbia Electric Carriage," *Town and Country*, September 15, 1906: 29; Ray Stannard Baker, "The Automobile in Common Use," *McClure's Magazine* 13, no. 3 (1899): 4.

70. "Electricity for Headlights," *Life* 56, no. 1458 (1910): 590.

71. "Electric Motor Vehicles," *Scientific American* 80, no. 19 (1899): 295.

72. Manhattan Electrical Supply Co., *Catalog of Automobile Supplies*.

73. "Annual Meeting of the Society of Automobile Engineers," *Horseless Age* 35, no. 3 (1915): 110–115.

74. Gelett Burgess, "Beating 'Em to It! The Manufacturers' Race for Supremacy," *Collier's* 50 (1912): 14.

75. Frances-Rand Company, "Safety for Fords," *Automobile Trade Journal* 22 (1909): 313.

76. "Signaling Devices," *Motor World Wholesale* 30 (1911): 337. See also Dean Electric Co., "Tuto Horn," *Literary Digest* 44 (1912): 1064.

77. "Display Ad 45," *New York Times*, January 2, 1910: AU12; "Wagner Electric Horn," 156; "Benjamin Auto Signals and the Button," *American Chauffeur: An Automobile Digest* 3 (1915): 275.

78. Cutler-Hammer Manufacturing Company, "Untitled," *Motor Age* 25 (1914): 4.

79. "Tuto Horn," 1064.

80. "The New Ever-Ready Electric Motor Car Horn," 1910 (Warshaw Collection of Business Americana, Archives Center, National Museum of American History, Smithsonian Institution, Washington, DC), 29.

81. "A One-Finger Emergency Automobile Brake," *Popular Scienec* 94, no. 4 (1919): 27. See also Roedding, "Push Button."

82. William Sparks, "Electric Push-Button," Patent No. 1,239,054, filed April 25, 1913; issued September 4, 1917.

83. "Turning the Car under the Garage Roof," *Motor World Wholesale* 37 (1913): 34.

84. Although not for automobiles specifically, household patents sought similar aims. See George Jepson, "Push Button," Patent No. 574,247, December 29, 1896; "The Crystal Palace Exhibition," *Electrical Engineer* 9, no. 7 (1892): 154.

85. "Handy Push Button for Auto Horns," *Popular Mechanics* 21, no. 3 (1914): 384.

86. Ray H. Manson, "Push-Button Switch," Patent No. 839,215, March 4, 1919.

87. Mary Dean on behalf of William Dean (deceased), "Push Button Switch," November 24, 1925. See also "Ignition Cut-Offs," *Horseless Age* 15 (1905): 696.

88. "Devices to Prevent Accident," *Atlanta Constitution*, March 14, 1904: 6.

89. "The Remote-Control or Push-Button Car," *Electric Railway Journal* 52, no. 3 (1918): 90.

90. "Push the Button," 63.

Chapter 8: Anyone Can Push a Button

1. "Electricity and Home Planning," *Illustrated World* 38, no. 2 (1922):261.

2. Arthur Brisbane, "The Littlest Woman in the World," *Cosmopolitan* 45 (1908): 328.

3. Walter Crane, *An Artist's Reminiscences* (New York: Macmillan, 1907), 379.

4. Harry C. Marillier, "The Automobile: A Forecast," *Eclectic Magazine of Foreign Literature, Science, and Art* 62, no. 6 (1895): 780.

5. "A City 100 Years from Now: The Luxuries People of the Next Century Will Probably Enjoy," *Washington Post*, March 5, 1911: LS5.

6. Frances Eldredge Russell, "Foundation and Fellowship," *Arena* 15 (1896): 962.

7. Regarding button "mania," see "Untitled," *Building Age* 15 (1893): 113.

8. J. H. Morgan, "Child Labor: Its Relation to Future Generations," in *Second Annual Report of the Department of Labor of the State of New York* (Albany, NY: The Argus Company, 1903), 575–577.

9. For histories of child labor in America, see Hugh D. Hindman, *The World of Child Labor: An Historical and Regional Survey* (Armonk, NY: M. E. Sharpe, 2009).

10. "Reports Containing the Cases Determined in All the Circuits from the United States Circuit Court of Appeals," 1907: 10–12.

11. "Circuit Court of Appeals," 10–12.

12. Carolyn Marvin, *When Old Technologies Were New: Thinking About Electric Communication in the Late Nineteenth Century* (New York: Oxford University Press, 1988); James R. Beniger, *The Control Revolution: Technological and Economic Origins of the Information Society* (Cambridge, MA: Harvard University Press, 1986).

13. David P. Parisi, "Tactile Modernity: On the Rationalization of Touch in the Nineteenth Century," in *Media, Technology, and Literature in the Nineteenth Century*, ed. Colette Colligan and Margaret Linley (Farnham, UK: Ashgate, 2011), 192.

14. Sigfried Giedion, *Mechanization Takes Command: A Contribution to Anonymous History* (Oxford: Oxford University Press, 1948).

15. Bryan S. Turner, *The Body and Society: Explorations in Social Theory* (Thousand Oaks, CA: Sage, 2008). Turner's ideas extend from Foucault, *Essential Works of Foucault.*

16. Turner, *The Body and Society.*

17. "Some Reaction-Time Studies," *Science* 6, no. 146 (1885): 458.

18. William Otterbein Krohn, *Practical Lessons in Psychology* (Chicago: The Werner Company, 1895).

19. Walter Moore Coleman, *Lessons in Hygienic Physiology* (New York: Macmillan, 1905).

20. M. V. O'Shea and John Harvey Kellogg, ed., *Making the Most of Life* (New York: Macmillan, 1915).

21. O'Shea and Kellogg, ed., *Making the Most of Life.*

22. The machine metaphor permeated many industries, especially medicine. See Samuel Osherson and Lorna AmaraSingham, "The Machine Metaphor in Medicine," in *Social Contexts of Health, Illness, and Patient Care,* ed. Elliot G. Mishler (Cambridge: Cambridge University Press, 1981), 228–229.

23. Frederick Winslow Taylor, *The Principles of Scientific Management* (New York: Harper & Brothers, 1913).

24. Taylor, *The Principles of Scientific Management.*

25. Angelo Mosso, *Fatigue,* trans. M. Drummond and W. B. Drummond (New York: G. P. Putnam's Sons, 1904); Frances Gulick Jewett, *Control of Body and Mind* (Boston: Ginn & Company, 1908). In the latter study, an experiment involved a boy and an electric button. When his hand was touched, a pendulum would begin to swing to measure the time it would take to press a button in response to the touch.

26. Bunting v. Oregon, 243 U.S. 426 (1917).

27. "High Mounting Saves Push Buttons on Grinders," *Electrical World* 78 (1921): 626; Louis Resnick, "Speeding up Production with Safety in a Small Plant," *National Safety News* 3–4 (1921): 4.

28. "3 Kicks in 6600 Repair Jobs," *Motor World Wholesale* 53, no. 10 (1917): 14.

29. James F. McLaughlin, "Push-Button," Patent No. 521,808, June 26, 1894: 26.

30. "Eighteenth Precinct Police Station," *Documents of the Senate of the State of New York* 31 (1917): 153.

31. "Accidents in Large Power Plants," *Western Electrician* 15 (1894): 259.

32. "Anchor Lock Push Button Switch," *Electrical Engineer* 26 (1898): 603. Another strategy for protecting against "malicious" use of push buttons involved altering the design of the button itself: Louis F. Johnson, "Electric Push-Button," Patent No. 631,892, 1899. Fire-alarm buttons were also used with malicious intent, thus necessitating a lock. See Max Straus, "Electric Push-Button," Patent No. 452,397, May 19, 1891. In boarding schools and orphanages, too, comings and goings were monitored via annunciator. See Hastings Hornell Hart, *Cottage and Congregate Institutions for Children* (New York: Charities Publication Committee, 1910). In other cases, homeowners were sometimes advised to make buttons "secret" so as to prevent inappropriate uses, particularly in the case of doorbells that were prone to "run-away rings"—presses carried out by pranksters who took advantage of the easily accessible button. More broadly, homeowners commonly incorporated "secret" spaces into the design of high end homes during this time period; these "off-stage" environments reflected how familial roles were performed and shed light on divides that might exist across class, gender, or race in domestic environs.

33. "Anchor Lock Push Button," 603.

34. E. L. Dunn, "Electric Dumb-Waiter Machines and Systems," *Electrical World* 50, no. 5 (1907): 223; "Fool Proof," *Natural Gas and Gasoline Journal* 13 (1919): 298; "'Fool-Proof' Electrical Apparatus," *Steel and*

Metal Digest 10 (1920): 469; "Fool-Proof Things," *Clay-worker* 77, no. 5 (1922): 485.

35. W. H. Leffingwell, "Increase Human Efficiency," *Natural Gas and Gasoline Journal* 13, no. 3 (1919): 190.

36. A. H. McIntire, "The Pushbutton Habit," *Electricity Journal* 14, no. 2 (1917): 42.

37. Lida Parce, *Economic Determinism: Or, the Economic Interpretation of History* (Chicago: Charles H. Kerr & Company, 1913), 152.

38. Parce, *Economic Determinism.*

39. "Self-Made Men," *Los Angeles Times*, March 31, 1925: A4.

40. Allan Louis Benson, "If Not Socialism—What?" *Truth about Socialism* 20 (1914): 168.

41. Benson, "If Not Socialism—What?"

42. "Daily Exercises for Millionaires," *Life* 79, no. 2057 (1922): 3.

43. Frank Dorrance Hopley, "The 'Push-the-Button Man,'" *Chicago Daily Tribune*, June 8, 1924: B6.

44. Hopley, "The 'Push-the-Button Man.'"

Chapter 9: Push for Your Pleasure

1. "Home a Hundred Years Hence: Some of the Wonderful Improvements People of the Next Century Will Enjoy," *Washington Post*, October 27, 1907: M2.

2. For a history of self-service, see Franck Cochoy, *On the Origins of Self-Service* (London: Routledge, 2015).

3. Alexandra Chasin, "Class and Its Close Relations: Identities among Women, Servants, and Machines," in *Posthuman Bodies*, ed. Judith M. Halberstam and Ira Livingston (Bloomington: Indiana University Press, 1995), 93.

4. "Calling the Servants," *New York Times*, April 7, 1929: SM10.

5. "Electricity," in *Compton's Pictured Encyclopedia* (Chicago: F. E. Compton & Company, 1922), 1114.

6. J. Salwyn Schapiro, *Modern and Contemporary European History* (Boston: Houghton Mifflin Company, 1918), 28.

7. "The Mystery of the Little Black Button," *Midwestern* 1, no. 1 (1906): 68–69. For further discussion of "magic" in discourses about technology, see William A. Stahl, "Venerating the Black Box: Magic in Media Discourse on Technology," *Science, Technology & Human Values* 20, no. 2 (1995): 234–258.

8. Alexander H. Robbins, "Reasonableness and Legal Right of Minimum Charge in Public Utilities," *Central Law Journal* 85, no. 6 (1916): 97.

9. "Picturing Electric Service to America's Millions," *N.E.L.A. Bulletin* 3, no. 8 (1916): 638.

10. "Prize-Winning Poster Selected for America's Electrical Week," *Electrical Review* 69, no. 8 (1916): 321.

11. Loring Pratt, "Untitled," *Union Electric Quarterly* 5, no. 9 (1917): 328.

12. Ruth Schwartz Cowan, *More Work for Mother: The Ironies of Household Technology from the Open Hearth to the Microwave* (New York: Basic Books, 1983), 228.

13. Judith Rollins, *Between Women: Domestics and Their Employers* (Philadelphia: Temple University Press, 1985), 208. See also Susan Leigh Star and Anselm Strauss, "Layers of Silence, Arenas of Voice: The Ecology of Visible and Invisible Work," *Computer Supported Cooperative Work* 8 (1999): 9–30.

14. Jacob Warshaw, *The New Latin America* (New York: Thomas Y. Crowell Company, 1922), 343.

15. Lundquist, "Electrical Goods in British," 121.

16. "Poster Advertising for Electrical Show," *Advertising & Selling* 30, no. 8 (1920): 31.

17. "Untitled," *Edison Monthly* 14 (1922): 43.

18. See Langdon Winner, *Autonomous Technology: Technics-out-of-Control as a Theme in Political Thought* (Cambridge, MA: MIT Press, 1977).

19. "The Public Servant," *Electrical World* 81 (1923): 11.

20. Lynn Harold Hough, "The Age of Machinery," *Michigan Technic* 33 (1920): 180.

21. "The Unseen Servant Behind the Perfect Meal Is the Perfect Refrigerator," *Literary Digest* 61 (1919): 146; "Electric Slave to Serve Homes," *Los Angeles Times*, May 18, 1932: A7.

22. Bristol Bell Co., "Advertisement 34—Untitled," *Scientific American* 80, no. 13 (1899): 208; Edison Company, "Philosophic Bits," 31.

23. For a discussion of the concept of "dream machines" as unrealized technologies that embodied social concerns of a time period, see Erkki Huhtamo, "Push the Button, Kinoautomat Will Do the Rest! Media-Archaeological Reflections on Audience Interactivity," in *Expanding Practices in Audiovisual Narrative*, ed. Raivo Kelomees and Chris Hales (Newcastle upon Tyne, UK: Cambridge Scholars Publishing, 2014).

24. W. B. Nesbit, "Push the Button," *Florida Star*, January 6, 1905: 2.

25. Nesbit, "Push the Button," 2.

26. Edward Morgan Forster, "The Machine Stops," in *Selected Stories*, ed. David Leavitt and Mark Mitchell (New York: Penguin Books, 2001), 94.

27. Marquis de Castellane, "Age of Little Effort," *Detroit Free Press*, November 22, 1903: D5.

28. Joseph French Johnson, "Advertising as an Economic Force," *Judicious Advertising* 18 (1920): 29.

29. Johnson, "Advertising as an Economic Force," 29.

30. "An Individually Operated Washing Machine Demonstration," *Merchandising Week* 18, no. 4 (1917): 209.

31. National Electric Light Association, "A Good Way to Sell Fans," *N.E.L.A. Bulletin* 4 (1917): 603.

32. "A Good Way to Sell Fans," 603.

34. "Press the Button," *American Hatter* 51 (1921): 92.

35. See also *How to Sell Electrical Labor-Saving*, 23–24.

36. "The Outer Adornment of Life," *Methodist Review* 76 (1894): 617–620.

37. Electrical Merchandising, ed. *How to Sell Electrical Labor-Saving Appliances: 119 Tested Plans for the Electric Store* (New York: McGraw-Hill, 1918).

38. John Cotton Dana, "Anticipations, or What We May Expect in Libraries," *Public Libraries* 12, no. 10 (1907): 382.

39. "City Directory Works Automatically," *Popular Science* 103, no. 1 (1923): 67; "Push-Button Map Tells Tourists How to Find Places in City," *Popular Mechanics* 43, no. 1 (1925): 81.

40. "The Informator," *Washington Post*, October 18, 1933: 6.

41. Anna Louise Strong, "Child-Welfare Exhibits," *Child-Welfare Exhibits: Types and Preparation* (Washington, DC: Government Printing Office, 1915), 41.

42. "Street Guide in Cars Aids Passengers and Crew," *Popular Mechanics* 41, no. 4 (1924): 555.

43. The World's Columbian Exposition was widely known for its illusions of spectacle and showmanship over education or substance. For specific analysis of the fair, see John G. Cawelti, "America on Display: The World's Fairs of 1876, 1893, 1933," in *The Age of Industrialism in America; Essays in Social Structure and Cultural Values*, ed. Frederic Cople Jaher (New York: Free Press, 1968), 317–363; Neil Harris, Wim de Wit, James Burkhart Gilbert, and Robert W. Rydell, and the Chicago Historical Society, *Grand Illusions: Chicago's World's Fair of 1893* (Chicago: Chicago Historical Society, 1993). Advertisers which emphasized that wiring one's home with electricity need not destroy the house or cost a

great deal. See Terrell Croft, *Wiring of Finished Buildings: A Practical Treatise Dealing with the Commercial and the Technical Phases of the Subject, for the Central-Station Man, Electrical Contractor and Wireman* (New York: McGraw-Hill, 1915).

44. William M. Brock, *Electricity in Paterson: Being a Treatise for Every-Day Folk on the Use of Electric Units and the Practical Application of Electric Currents* (Paterson, NJ: Press of the Sunday Chronicle, 1896; collections of the Huntington Library, Pasadena, CA), 52.

45. Martin Hussobee, "Electric Light Companies Appeal to Public in Big Campaign," *Printers' Ink* 114 (1921): 10–17.

46. George F. Oxley, "Movie Program Making Progress," *N.E.L.A. Bulletin* 8, no. 11 (1921): 660.

47. George F. Oxley, "More Than Two Million People View First N.E.L.A. Film—New Picture Is Shown at Convention," *N.E.L.A. Bulletin* 9, no. 6 (1922): 357.

48. Sidney Arnold, "Random Notes and Sketches," *American Artisan and Hardware Record* 85 (1923): 16.

49. Preston S. Arkwright, "Shows Why Services Worth Millions Cannot Be Given to Consumers Free," *Public Service Management* 32, no. 1 (1922): 55.

50. Alfred Kuttner, "Freud's Contribution to Ethics," *New Republic* 5 (1915): 101.

Chapter 10: Conclusion

1. Hughes, "The Woman's Slant," *Life* 99 (1932): 42.

2. Charles G. Armstrong, "Improvements in Annunciator and Bell Work," *Electrical Engineer* 12 (1891): 685.

3. J. Elliott Shaw Company, *Household and Experimental Electrical Supplies and Novelties, Catalog No. 15*, 1903 (Warshaw Collection of Business Americana, Archives Center, National Museum of American History, Smithsonian Institution, Washington, DC).

4. Thomas Hine, *Populuxe* (New York: Knopf, 1986), 128. Hine notes that in the 1950s, switches other than push buttons carried out tasks just as well, if not better than buttons, but they were considered fashionable and thus the best interface choice. He writes, "The significance of the push button was entirely symbolic. It had merely replaced another kind of control device that had been doing the job satisfactorily for many years. In most cases push buttons replaced thermostat dials, which were not merely as good as push buttons for most applications but quite a lot better, since they allowed even more subtle and exact adjustment than could hundreds of push buttons."

5. This relates to a discussion that has occurred in STS literature about invisible infrastructure. Interfaces often play a role in hiding infrastructure. See, for example, Geoffrey C. Bowker and Susan Leigh Star, *Sorting Things Out: Classification and Its Consequences* (Cambridge, MA: MIT Press, 1999); Geoffrey C. Bowker and Susan Leigh Star, "How to Infrastructure," in *Handbook of New Media*, ed. Leah A. Lievrouw and Sonia Livingstone (Thousand Oaks, CA: Sage, 2002), 230–245; S. L. Star and K. Ruhleder, "Infrastructures Are Also Relational: Steps toward an Ecology of Infrastructure: Design and Access for Large Information Spaces," *Information Systems Research* 7, no. 1 (1996): 111–134.

6. Hine, *Populuxe*.

7. Alfred Leech, "Your New Home of 1980 May Operate by Thumb," *Washington Post*, December 8, 1956: C4.

8. ". . . Closer Than We Think!" *Chicago Daily Tribune*, October 30, 1960: F8.

9. Numerous scholars have written on this issue. Some examples include Ruth Schwartz Cowan, *More Work for Mother: The Ironies of Household Technology from the Open Hearth to the Microwave* (New York: Basic Books, 1983); Susan Strasser, *Never Done: A History of American Housework* (New York: Pantheon Books, 1982); Christina Hardyment, *From Mangle to Microwave: The Mechanization of Household Work* (Cambridge: Polity Press, 1988); Davin Heckman, *A Small World: Smart Houses and the Dream of the Perfect Day* (Durham, NC: Duke University Press, 2008).

10. De Soto, "Car That Has All Its Buttons," *Life* 45, no. 25 (1959): 11.

11. "Foresees Age of Push Button for Housewife," *Chicago Daily Tribune*, January 13, 1950: A1.

12. Jessie Cartwright, "You Just Push a Button? O, Yeah!" *Chicago Daily Tribune*, March 9, 1958: E4.

13. Bill DeRouchey, "1963 Push Button Phone," December 10, 2006, video, 0:31, https://www.youtube.com/watch?v=5t5na44D0Dw.

14. DeRouchey, "1963 Push Button Phone."

15. Zenith Radio Corporation, "Why You Can Operate Zenith TV from Your Easy Chair," *Coronet*, February 1951 (retrieved from Duke Ad*Access).

16. Lloyd Norman, "If War Comes It'll Be Long, Brucker Says," *Chicago Daily Tribune*, February 2, 1956: 12.

17. See, for example, Joseph Alsop and Stewart Alsop, "Are We Ready for Push-Button War?" *Saturday Evening Post* 220, no. 10 (1947): 18–104; John G. Norris, "Calamity Howlers and Pollyannas Both Wrong on 'Push-Button War,'" *Washington Post*, August 3, 1947: B11; Cabell Phillips, "Why We're Not Fighting with Push Buttons," *New York Times*, July 16, 1950: SM7.

18. "A Plea for Individualism: Noted Architect Describes Failure of Educational, Political and Economic Systems in Bicentennial Conference Address," *Princeton Alumni Weekly* 47 (1946): 6.

19. Ben Prawdzik, "Milgram Experiment, 50 Years On," *Yale Daily News*, September 28, 2011, http://yaledailynews.com/blog/2011/09/28/milgram-experiment-50-years-on/.

20. For one take on the experiment, see Michael Shermer, "What Milgram's Shock Experiments Really Mean," http://www.scientificamerican.com/article.cfm?id=what-milgrams-shock-experiments-really-mean/.

21. Jack Geyer, "The Pillow Pilots of the Push-Button Age," *Los Angeles Times*, September 24, 1952: A5.

22. John H. Averill, "Underexercised Nation," *Los Angeles Times*, December 6, 1961: C1.

23. "Even Crossing Street Can Be Big Problem in Push-Button World," *Los Angeles Times*, December 10, 1961: WS5.

24. Douglas Engelbart, quoted in Thierry Bardini, *Bootstrapping: Douglas Engelbart, Coevolution, and the Origins of Personal Computing* (Stanford, CA: Stanford University Press, 2000), 99.

25. Engelbart in Bardini, *Bootstrapping*, 101.

26. Logitech, "Introducing the Most Agile Mouse Ever to Set Foot on a Desktop," *Byte* (1998): 1.

27. Soren Pold, "Button," in *Software Studies: A Lexicon*, ed. Matthew Fuller (Cambridge, MA: MIT Press, 2008), 32.

28. This refers to the concept of "remediation." For further discussion, see Jay David Bolter and Richard Grusin, *Remediation: Understanding New Media* (Cambridge, MA: MIT Press, 1998).

29. Wendy Hall, "Ending the Tyranny of the Button," *IEEE MultiMedia* 1, no. 1 (1994): 60.

30. Pold, "Button," 35.

31. Pold, 34.

32. Marita Sturken, Douglas Thomas, and Sandra Ball-Rokeach, *Technological Visions: The Hopes and Fears That Shape New Technologies* (Philadelphia: Temple University Press, 2004).

33. Michel Foucault, *Power/Knowledge: Selected Interviews and Other Writings 1972–1977* (New York: Pantheon Books, 1980), 58.

34. For Ninette, see https://youtu.be/tdJVj2x9wNk. The song describes the "world gone mad" from so many automatic, push-button devices and ends with an allusion to push-button warfare and nuclear destruction. Money Mark's techno song portrays a high-tech version of pushes and clicks: https://youtu.be/oJDGcxAf9D8; Sugababes's song offers a thinly veiled sexual metaphor. The women encourage a man with the

lyrics, "If you're ready for me boy you'd better push the button and let me know." The music video takes place on an elevator. See https://youtu.be/B-qTCrNpi44.

35. Jeff Woolnough, dir., *The Outer Limits*, season 3, episode 13, "Dead Man's Switch," aired April 4, 1997, on Showtime. Regarding original material for the film *The Box*, see Richard Matheson, "Button, Button," in *Collected Stories*, vol. 3, ed. Stanley Wiater (Colorado Springs, CO: Gauntlet Press, 2005). The short story, originally published in *Playboy*, inspired a popular episode of *The Twilight Zone*. See Peter Medak, dir., *The Twilight Zone*, season 1, episode 50, "Button, Button," aired March 7, 1985, on CBS.

36. Rob Walker, "Ad Play," *New York Times Magazine*, December 17, 2006, http://www.nytimes.com/2006/12/17/magazine/17wwln_consumed.t.html.

37. Walker, "Ad Play."

38. Eliot Phillips, "Staples Easy Button Hacking," *Hackaday*, July 18, 2006, http://www.hackaday.com/2006/07/18/staples-easy-button-hacking/.

39. Phillips, "Staples Easy Button Hacking."

40. See https://www.reddit.com/r/thebutton/.

41. Kate Knibbs, "People Got So into This Strange Internet Button They Made Up Religions," *Gizmodo*, May 12, 2015, https://gizmodo.com/people-got-so-into-this-strange-internet-button-they-ma-1700779699/.

42. Klint Finley, "Press Amazon's IRL Button to Call a Cab, Brew Coffee, or Whatever," *Wired*, May 13, 2016, http://www.wired.com/2016/05/aws-iot-dash/.

43. Des Traynor, "On Magical Software," *Inside Intercom* (blog), July 2, 2014, http://blog.intercom.io/on-magical-software/.

44. Lars-Erik Janlert, "The Ubiquitous Button," *Interaction* 21, no. 3 (2014): 26.

45. Janlert, "The Ubiquitous Button."

46. Siegler, "Title."

47. Kyle Vanhemert, "Apple's Haptic Tech Is a Glimpse at the UI of the Future," *Wired*, March 19, 2015, http://www.wired.com/2015/03/apples-haptic-tech-makes-way-tomorrows-touchable-uis/, accessed January 15, 2016. See also Megan Fellman, "Bringing Texture to Your Flat Touchscreen," *Northwestern University News*, February 9, 2015, http://www.northwestern.edu/newscenter/stories/2015/02/bringing-texture-to-your-flat-touchscreen.html, accessed March 31, 2016.

48. Damon Lavrinc, "Apple's New Trackpad Might Be the Solution to the Touchscreen Problem," *Jalopnik*, March 24, 2015, http://jalopnik.com/apple-s-new-trackpad-might-be-the-solution-to-the-touch-1693383080, accessed January 15, 2016.

49. Bob Sorokanich, "Bosch Is Working on Touchscreen Buttons You Can Actually Feel," *Road & Track*, November 11, 2015, http://www.roadandtrack.com/new-cars/car-technology/news/a27320/bosch-is-working-on-touchscreen-buttons-you-can-actually-feel/, accessed May 1, 2016.

50. Marquis de Castellane, "An Age of Little Effort," *Detroit Free Press*, November 22, 1903: D5.

51. J. R. Parker, "Buttons, Simplicity, and Natural Interfaces," *Loading ...* 2, no. 2 (2008). http://journals.sfu.ca/loading/index.php/loading/article/view/33/.

52. Russell Brandom, "Death of the Button," *Kempt*, September 30, 2009, http://www.getkempt.com/gadgetry/the-death-of-the-button.php.

53. Jennifer Hoar, "Apple's Steve Jobs Hates Buttons," *CBS News*, July 25, 2007. http://www.cbsnews.com/news/apples-steve-jobs-hates-buttons/.

54. Adam Clark Estes, "Steve Jobs Didn't Want Apple Devices to Have an Off Switch," *Wire*, October 24, 2011, http://www.thewire.com/technology/2011/10/isaacson-jobs-60-minutes-interview/44028/.

55. P. Jackson and J. Lethem, eds., *The Exegesis of Philip K. Dick* (Boston: Houghton Mifflin Harcourt, 2011), 499.

56. See Laura Sydell, "Microsoft's Kinect Brings Gesture to a New Level," *NPR*, November 4, 2010, http://www.npr.org/templates/story/story.php?storyId=131074438; Keith Wagstaff, "Project Glass: Google's Augmented Reality Glasses Take Hands-Free Computing to the Extreme," *Time*, April 4, 2012. http://techland.time.com/2012/04/04/googles-augmented-reality-glasses-take-hands-free-computing-to-the-extreme/.

57. Martin Dodge and Rob Kitchin, "Towards Touch-Free Spaces: Sensors, Software and the Automatic Production of Shared Public Toilet," in *Touching Space, Placing Touch*, ed. Mark Paterson and M. Dodge (Burlington, VT: Ashgate, 2012).

58. Lev Manovich, *The Language of New Media* (Cambridge, MA: MIT Press, 2001).

59. Will Greenwald, "Hands on with the Oculus Rift and Oculus Touch Controllers," *PC Magazine*, January 7, 2016, https://www.pcmag.com/article2/0,2817,2497577,00.asp.

60. Tim Wu, "The Problem with Easy Technology," *New Yorker*, February 21, 2014, http://www.newyorker.com/tech/elements/the-problem-with-easy-technology/.

61. Quoted in Ramin Setoodeh and Elizabeth Wagmeister, "Matt Lauer Accused of Harassment by Multiple Women," *Variety*, November 29, 2017, http://variety.com/2017/biz/news/matt-lauer-accused-sexual-harassment-multiple-women-1202625959/.

62. Megh Wright, "Seth Meyers Takes 'A Closer Look' at the Matt Lauer Sexual Misconduct Allegations," *SplitSider*, December 1, 2017. http://splitsider.com/2017/12/seth-meyers-takes-a-closer-look-at-the-matt-lauer-sexual-misconduct-allegations/.

63. Paul Farhi, "So, You Had Questions about Matt Lauer's Desk?" *Washington Post*, December 1, 2017, https://www.washingtonpost.com/lifestyle/style/so-you-had-questions-about-that-button-on-matt-lauers-desk/2017/12/01/48b1f7c2-d6bd-11e7-a986-d0a9770d9a3e_story.html.

64. Donald J. Trump (@realDonaldTrump), Twitter, January 2, 2018, 4:49 p.m., https://twitter.com/realdonaldtrump/status/948355557022420992.

65. David Emery, "Is President Trump's Nuclear Button Bigger Than Kim Jong-un's?" *Snopes*, January 3, 2018, https://www.snopes.com/2018/01/03/nuclear-button-trump-kim/.

66. Michael T. Klare, "Whose Finger on the Nuclear Button?" *Common Dreams*, November 6, 2016, https://www.commondreams.org/views/2016/11/06/whose-finger-nuclear-button

67. Emery, "President Trump's Nuclear Button."

Further Reading

1. Marquis de Castellane, "An Age of Little Effort," *Detroit Free Press*, November 22, 1903: D5.

2. James R. Beniger, *The Control Revolution: Technological and Economic Origins of the Information Society* (Cambridge, MA: Harvard University Press, 1986).

3. Carolyn Marvin, *When Old Technologies Were New: Thinking About Electric Communication in the Late Nineteenth Century* (New York: Oxford University Press, 1988).

4. Thomas Park Hughes, *Networks of Power: Electrification in Western Society, 1880–1930* (Baltimore: Johns Hopkins University Press, 1983).

5. David E. Nye, *Electrifying America: Social Meanings of a New Technology, 1880–1940* (Cambridge, MA: MIT Press, 1990).

6. Chris Otter, *The Victorian Eye: A Political History of Light and Vision in Britain, 1800–1910* (Chicago: University of Chicago Press, 2008). This text, however, focuses on switches primarily from a visual perspective.

7. Ruth Schwartz Cowan, "The 'Industrial Revolution' in the Home: Household Technology and Social Change in the 20th Century," *Technology and Culture* 17, no. 1 (1976): 1–23.

8. See Ruth Schwartz Cowan, *More Work for Mother: The Ironies of Household Technology from the Open Hearth to the Microwave* (New York: Basic Books, 1983); Susan Strasser, *Never Done: A History of American Housework* (New York: Pantheon Books, 1982).

9. Davin Heckman, *Small World: Smart Houses and the Dream of the Perfect Day* (Durham, NC: Duke University Press, 2008).

10. For examples of these histories, see Claude S. Fischer, *America Calling: A Social History of the Telephone to 1940* (Berkeley: University of California Press, 1992); Marvin, *When Old Technologies Were New*; Tom Standage, *The Victorian Internet: The Remarkable Story of the Telegraph and the Nineteenth Century's On-Line Pioneers* (New York: Walker and Co., 1998).

11. For a history of changes to internal communication practices in businesses, see Joanne Yates, *Control through Communication: The Rise of System in American Management* (Baltimore: Johns Hopkins University Press, 1989).

12. "Electric Current-Selling Devices," *Electrical Record and Buyer's Reference* 5, no. 6 (1909): 262.

13. Jean Baudrillard, *The System of Objects* (London: Verso, 2005), 58–59.

14. Michel de Certeau, Luce Giard, and Pierre Mayol, *The Practice of Everyday Life: Living and Cooking*, vol. 2 (Minneapolis: University of Minnesota Press, 1998), 212.

15. de Certeau et al., *The Practice of Everyday Life*, 212.

16. Langdon Winner, *Autonomous Technology: Technics-out-of-Control as a Theme in Political Thought* (Cambridge, MA: MIT Press, 1977), 42.

17. Vilém Flusser, *Shape of Things: A Philosophy of Design* (London: Reaktion Books, 1999), 92.

18. This approach extends from Pablo Boczkowski and Ignacio Siles, "Steps toward Cosmopolitanism in the Study of Media Technologies: Integrating Scholarship on Production, Consumption, Materiality, and Content," in *Media Technologies: Essays on Communication, Materiality, and Society*, ed. Tarleton Gillespie, Pablo Boczkowski, and Kirsten A. Foot (Cambridge, MA: MIT Press, 2014), 53–76.

References

Adams, Joseph Henry, and Joseph B. Baker. *Harper's Electricity Book for Boys*. New York: Harper & Brothers, 1907.

Adams, Judith A. "The Promotion of New Technology through Fun and Spectacle: Electricity at the World's Columbian Exposition." *Journal of American Culture* 18, no. 2 (1995): 45–55.

Adams-Morgan Co. *Electrical Apparatus and Supplies: Dynamos, Motors, Chemicals, Wireless Telegraph and Telephone Instruments, Books and Tools*. Upper Montclair, NJ: Adams-Morgan, ca. 1911. Collections of the Bakken Museum, Minneapolis, MN.

A. J. Wilkinson Co. *Illustrated Catalogue of Electrical Goods and Bell-Hangers Supplies*. 1891. Warshaw Collection of Business Americana. Archives Center, National Museum of American History, Smithsonian Institution, Washington, DC.

A. J. S. "Turn-out-the-Light." *Rotarian*, November 1924: 51–52.

Alden, W. L. "Life's Little Worries." *Pearson's Magazine* 5 (1898): 558–561.

Allen, John, and Chris Hamnett. *A Shrinking World? Global Unevenness and Inequality*. Oxford: Open University and Oxford University Press, 1995.

Allison, William L. *Allison's Webster's Counting-house Dictionary of the English Language and Dictionary of Electricity, Electrical Terms and Apparatus*. New York: William L. Allison, 1886.

Allsop, Frederick Charles. *Practical Electric Bell Fitting*. London: E & F. N. Spon, Ltd., 1907.

Allsop, F. C. "Electrical Apparatus, Constructing and Repairing.—XXX." *English Mechanic and World of Science and Art* 60 (1894): 149.

Alsop, Joseph, and Stewart Alsop. "Are We Ready for Push-Button War?" *Saturday Evening Post* 220, no. 10 (1947): 18–104.

The American Institute of Electrical Engineers. *The Boston Electrical Handbook: Being a Guide for Visitors from Abroad Attending the International Electrical Congress, St. Louis, Mo.* Boston: American Institute of Electrical Engineers, 1904. Collections of the Bakken Museum, Minneapolis, MN.

"American Notes." *Electricity Journal* 21 (1888): 185–186.

American Supply Company. "Electric Button." *Popular Mechanics*, March 1906.

Anthony, Stanley. "Push Button." Patent No. 1,340,139. Filed March 19, 1919; issued May 18, 1920.

Architectural File. "'H&H' Automatic Door Switches." 1920: 1602.

Ardee Manufacturing Co. *Ardee Manufacturing Co. Illustrated Catalogue: Manufacturers, Importers and Jobbers of Toys, Novelties and Mail Order Merchandise*. Stamford, CT: Ardee Manufacturing Co., ca. 1903. Collections of the Baker Library, Harvard Business School, Boston, MA.

Arkwright, Preston S. "Shows Why Services Worth Millions Cannot Be Given to Consumers Free." *Public Service Management* 32, no. 1 (1922): 55–56.

Armstrong, Charles G. "Improvements in Annunciator and Bell Work." *Electrical Engineer* 12 (1891): 685.

Arnold, Sidney. "Random Notes and Sketches." *American Artisan and Hardware Record* 85 (1923): 16.

The Arrester. "Foreign Currents." *American Telephone Journal* 7 (1903): 211–212.

Atchison Daily Globe. "The Widow of President Polk: She Touched the Electric Button and Started the Cincinnati Exposition." July 21, 1888.

Atkinson, W. W. "Push Versus Pull." *Santa Fe Employees' Magazine* 3 (1909): 1017–1018.

Atlanta Constitution. "A Boy and a Bell." November 4, 1900: B2.

Atlanta Constitution. "Devices to Prevent Accident." March 14, 1904: 6.

Atlanta Constitution. "The Electric Call Bells." May 1, 1885.

Atlanta Constitution. "M'Coy Wants Money: The Private Detective Claims Cash for Spotting Gamblers." March 14, 1896: 5.

Atlanta Constitution. "To Press the Button: President Cleveland, at Gray Gables, Will Start Our Exposition." July 23, 1895: 11.

"At the Front Door." *Century: A Popular Quarterly* 5 (1873): 509.

Autosales Gum & Chocolate. "One at Every Station." 1912. Warshaw Collection of Business Americana. Archives Center, National Museum of American History, Smithsonian Institution, Washington, DC.

Averill, John H. "Underexercised Nation." *Los Angeles Times*, December 6, 1961.

Badt, F. B. *Bell-Hangers' Hand Book.* Chicago: Electrician Publishing Company, 1889.

Baily, E. A. "President Ballard Addresses Baltimore Section." *N.E.L.A. Bulletin* 7, no. 2 (1920): 135–136.

Baker, Ray Stannard. "The Automobile in Common Use." *McClure's Magazine* 13, no. 3 (1899): 4.

Bakhtin, Mikhail. *Rabelais and His World.* Translated by H. Iswolsky. Bloomington: Indiana University Press, 1993.

Bardini, Thierry. *Bootstrapping: Douglas Engelbart, Coevolution, and the Origins of Personal Computing.* Stanford, CA: Stanford University Press, 2000.

Barnard, Charles. "Some Queer Houses." *Youth's Companion* 65, no. 51 (1892): 2.

Batchelder, Charles Clarence. "The Grain of Truth in the Bushel of Christian Science Chaff." *Popular Science* 72, no. 13 (1908): 211–223.

Baudrillard, Jean. *The System of Objects*. London: Verso, 2005.

Baum, L. Frank. *The Master Key: An Electrical Fairy Tale*. Indianapolis, IN: Bowen-Merrill, 1901.

Baum, William M. "Address of Welcome." *Annual Report of the Michigan Dairymen's Association* 23 (1907): 21–23.

Bax, Edmund Ironside. *Popular Electric Lighting: Being Practical Hints to Present and Intending Users of Electric Energy for Illuminating Purposes With a Chapter on Electric Motors*. London: Biggs & Co., 1891.

Beach, Alfred E., ed. "The Miniature Telegraph." In *The Science Record for 1874: A Compendium of Scientific Progress and Discovery During the Past Year*, 111–113. New York: Munn & Company, 1874.

Beach, Robin, and Ernest John Streubel. *Electricity and Magnetism: The Science of Power*. Vol. 2. New York: P.F. Collier & Son Company, 1922.

Bedford-Jones, H. "The Problem Answerers." *Popular Electricity and the World's Advance* 6 (1913): 146–151.

Beebe, H. "Physiology: The Organic Nervous System." *Medical Century: An International Journal of Homeopathic Medicine and Surgery* 6, no. 7 (1898): 207–212.

Beers, A. Henry, Jr. "Winning the Architect Instead of 'Forcing' Him: Why the Western Electric Co. Revised Its First Advertising of Habirshaw House Wiring." *Printers' Ink* 102 (1918): 40–46.

Beeton, Isabella Mary. *Beeton's Housekeeper's Guide; Comprising Complete and Practical Instructions on House Building, Buying, and Furnishing; the Decoration of the Home; the Economical Management of the Household; and the Treatment of Children in Health and Sickness*. London: Ward, Lock, and Co., ca. 1890.

Bellamy, Edward. *Equality*. New York: D. Appleton and Company, 1897.

Beniger, James R. *The Control Revolution: Technological and Economic Origins of the Information Society*. Cambridge, MA: Harvard University Press, 1986.

Benjamin, Walter. "On Some Motifs in Baudelaire." In *Illuminations: Essays and Reflections*, edited by Hannah Arendt. New York: Schocken, 1968.

Bennett, Reginald A. R. "Electrical Bells: How to Make and Use Them." *Boy's Own Annual* 15 (1893): 717.

Bennett, Reginald A. R. *How to Make Electrical Machines*. New York: Frank Tousey, 1902.

Benson, Allan Louis. "If Not Socialism—What?" *Truth about Socialism* 20 (1914): 168.

Benson, James W. *Time and Time-Tellers*. London: Robert Hardwicke, 1875.

Bernard, Andreas. *Lifted: A Cultural History of the Elevator*. New York: New York University Press, 2014.

Betcone, David S. *The Aladdin Builders' Safeguard; Containing Practical Methods of Excavating, Laying Footings, Walks, Piers, Building Foundations, Chimneys, Etc., Installing Heating, Plumbing and Wiring*. Bay City, MI: North American Construction Co., 1914. Collections of the Winterthur Library, Winterthur, DE.

Bidwell, George. *Forging His Own Chains: The Wonderful Life-Story of George Bidwell*. Hartford, CT: The Bidwell Publishing Company, 1890.

Boczkowski, Pablo, and Ignacio Siles. "Steps toward Cosmopolitanism in the Study of Media Technologies: Integrating Scholarship on Production, Consumption, Materiality, and Content." In *Media Technologies: Essays on Communication, Materiality, and Society*, edited by Tarleton Gillespie, Pablo Boczkowski, and Kirsten A. Foot, 53–76. Cambridge, MA: MIT Press, 2014.

Boston Daily Globe. "About Those Buttons: One Man Who Saw How They Didn't Work Very Well." July 23, 1895.

Boston Daily Globe. "The Automatic Fire Alarm Button." November 19, 1899.

Boston Daily Globe. "A Call-Bell Wrinkle." June 20, 1890: 2.

Boston Daily Globe. "Editorial Points." September 18, 1898.

Boston Daily Globe. "Fair Opened in a Blaze of Glory." May 1, 1904: A7.

Boston Daily Globe. "Great Inventor Hardly Known: Stephen Dudley Field, Man of Wonders." April 17, 1910: SM3.

Boston Daily Globe. "He Presses the Button." March 18, 1895: 10.

Boston Daily Globe. "Push Button in Evidence." February 15, 1895: 12.

Boston Daily Globe. "Pushed Button in Vain." August 19, 1906: SM2.

Boston Daily Globe. "Thought He Was on Pullman; Traveler on the Street Car Confuses the Use of Push Buttons." June 21, 1906: SM16.

Boston Daily Globe. "To Attract Page's Attention." January 10, 1900.

Bottone, Selimo Romeo. *Electric Bells and All About Them: A Practical Book for Practical Men.* London: Whittaker & Co., 1889.

Bowditch, H. P., and Wm. F. Southard. "A Comparison of Sight and Touch." *Journal of Physiology* 3 (1882): 232–245.

Bowker, Geoffrey C., and Susan Leigh Star. "How to Infrastructure." In *Handbook of New Media*, edited by Leah A. Lievrouw and Sonia Livingstone, 230–245. Thousand Oaks, CA: Sage, 2002.

Bowker, Geoffrey C., and Susan Leigh Star. *Sorting Things Out: Classification and Its Consequences.* Cambridge, MA: MIT Press, 1999.

Brandom, Russell. "The Death of the Button." *Kempt*, September 30, 2009. http://www.getkempt.com/gadgetry/the-death-of-the-button.php.

Brisbane, Arthur. "The Littlest Woman in the World." *Cosmopolitan* 45 (1908): 324–329.

Bristol Bell Co. "Advertisement 34—Untitled." *Scientific American* 80, no. 13 (1899): 208.

Brock, William M. *Electricity in Paterson: Being a Treatise for Every-Day Folk on the Use of Electric Units and the Practical Application of Electric Currents.* Paterson, NJ: Press of the Sunday Chronicle, 1896. Collections of the Huntington Library, San Marino, CA.

Brody, Sidney. "Pepper ... and Salt." *Wall Street Journal*, January 14, 1965: 14.

Brooks, H. B. *Electrical Instruments in England. Special Agents Series—No. 55.* Washington, DC: Government Printing Office, 1912.

Brown, H. C. "At the Home of the Kodak." *Harper's New Monthly Magazine* 83 (1891): 17–22.

Brown and Murray Co. "Our Experience." ca. 1914. Warshaw Collection of Business Americana. Archives Center, National Museum of American History, Smithsonian Institution, Washington, DC.

Brusseau, Edward. "Push-Button." Patent No. 780,860. January 24, 1905.

Bryant Electric Company. *Electrical Wiring Specifications for Residences and Apartment Houses.* Bridgeport, CT: Bryant Electric Company, 1914.

Buchanan, John F. *Brassfounders' Alloys: A Practical Handbook Containing Many Useful Tables, Notes and Data, for the Guidance of Manufacturers and Tradesmen.* London: E. & F. N. Spon, Ltd., 1901.

Bullard, William Hannum Grubb. *Naval Electricians' Text and Handbook.* Annapolis, MD: US Naval Institute, 1904.

Bunting v. Oregon, 243 U.S. 426 (1917).

Burgess, Gelett. "Beating 'Em to It! The Manufacturers' Race for Supremacy." *Collier's* 50 (1912): 14.

Burns, Tracy Walling, and Julius Martin. *Electrical Installations of the United States Navy: A Manual of the Latest Approved Material, Including Use, Operation, Inspection, Care and Management and Method of Installation on Board Ship.* Annapolis, MD: US Naval Institute, 1907.

Caldwell, Otis W. "Natural History in the Grades." *Elementary School Teacher* 11, no. 2 (1910): 49–62.

"The Camerist." *American Amateur Photographer* 8, no. 5 (1896): 224.

Candee, Helen Churchill. "House Building." In *The House and Home: A Practical Book*, edited by Lyman Abbott. New York: Charles Scribner's Sons, 1896.

Carnegie, Andrew. *Triumphant Democracy: Or, Fifty Years' March of the Republic.* New York: Charles Scribner's Sons, 1886.

Cartoons Magazine. "Thinking in Terms of—." 1918: 51.

Cartwright, Jessie. "You Just Push a Button? O, Yeah!" *Chicago Daily Tribune,* March 9, 1958: E4.

Casson, Herbert Newton. *The History of the Telephone.* Chicago: A. C. McClurg & Co., 1910.

Castellane, Marquis de. "An Age of Little Effort." *Detroit Free Press,* November 22, 1903: D5.

Cawelti, John G. "America on Display: The World's Fairs of 1876, 1893, 1933." In *The Age of Industrialism in America; Essays in Social Structure and Cultural Values,* edited by Frederic Cople Jaher, 317–363. New York: Free Press, 1968.

Chasin, Alexandra. "Class and Its Close Relations: Identities among Women, Servants, and Machines." In *Posthuman Bodies,* edited by Judith M. Halberstam and Ira Livingston, 73–96. Bloomington: Indiana University Press, 1995.

Chicago Daily Tribune. "The Age of Buttons." June 8, 1900.

Chicago Daily Tribune. "Baby Marion May Press the Button: Little Daughter of President Cleveland to Start Atlanta's Exposition." July 23, 1895.

Chicago Daily Tribune. "... Closer Than We Think!" October 30, 1960: F8.

Chicago Daily Tribune. "Foresees Age of Push Button for Housewife." January 13, 1950: A1.

Chicago Daily Tribune. "Odd Phases of Chicago Life as Presented to an Observer in a Single Day." May 4, 1896: 1.

Chicago Daily Tribune. "Oppose Push Buttons on Cars." July 15, 1900: 14.

Chicago Daily Tribune. "Pullman's New Double Decker Street Car." November 6, 1897: 7.

Chicago Daily Tribune. "Push Buttons and Thermometers." December 5, 1907: 10.

Chicago Daily Tribune. "Slot Machines to Replace Department Store Clerks." May 15, 1910: 23.

Child, Charles Tripler. *The How and Why of Electricity: A Book of Information for Non-Technical Readers.* New York: D. Van Nostrand Company, 1905.

C. H. W. Bates & Co. "Electric Chestnut Bell." In *C.H.W. Bates & Co. Big Bargain Catalog.* Boston: C. H. W. Bates & Co., n.d. Warshaw Collection of Business Americana. Archives Center, National Museum of American History, Smithsonian Institution, Washington, DC.

Clark, T. M. *The Care of a House; a Volume of Suggestions to Householders, Housekeepers, Landlords, Tenants, Trustees, and Others, for the Economical and Efficient Care of Dwelling-Houses.* New York: Macmillan, 1903.

Classen, Constance. *The Deepest Sense: A Cultural History of Touch.* Champaign: University of Illinois Press, 2012.

Claudy, C. H. "Wiring for Electric Lighting." *Suburban Life, the Countryside Magazine* 19, no. 3 (1914): 140–144.

Clewell, Clarence Edward. *Handbook of Machine Shop Electricity.* New York: McGraw-Hill, 1916.

"Clocks Which Furnish Light." In *Our Wonderful Progress: The World's Triumphant Knowledge and Works,* edited by Trumbull White, 303–305. Springfield, MA: Hampden Publishing Company, 1902.

Cochoy, Franck. *On the Origins of Self-Service.* London: Routledge, 2015.

Cohn, F. W. "Electrical Push Button." Patent No. 859,367. July 9, 1907.

Coleman, Grace D. "The Efficiency of Touch and Smell." *American Annals of the Deaf* 67 (1922): 301–325.

Coleman, Walter Moore. *Lessons in Hygienic Physiology.* New York: Macmillan, 1905.

Colligan, Colette, and Margaret Linley. *Media, Technology, and Literature in the Nineteenth Century: Image, Sound, Touch*. Farnham, UK: Ashgate, 2011.

Collins, A. F. "Electricity around the House." *Boys' Life* 6, no. 6 (1916): 32.

Comstock, William T. *Two-Family and Twin Houses*. New York: William T. Comstock, 1908.

Cooke, Augustus Paul. "Training the College Mind toward G.E." *Printers' Ink* 112, no. 12 (1920): 89.

Country Life in America. "Western Electric Inter-Phone." 1913: 101.

Countryside Magazine and Suburban Life. "Western Electric Inter-Phone." 1914: 221.

Cowan, Ruth Schwartz. *More Work for Mother: The Ironies of Household Technology from the Open Hearth to the Microwave*. New York: Basic Books, 1983.

Cowan, Ruth Schwartz. "The 'Industrial Revolution' in the Home: Household Technology and Social Change in the 20th Century." *Technology and Culture* 17, no. 1 (1976): 1–23.

Crane, Walter. *An Artist's Reminiscences*. New York: Macmillan, 1907.

Cranny-Francis, Anne. *Technology and Touch: The Biopolitics of Emerging Technologies*. New York: Palgrave Macmillan, 2013.

Crans, E. G. "Luxury in Modern Living." *Puritan* 5 (1899): 218–228.

Crecelius, P. "Repairing the Electric Bell." In *The Twentieth Yearbook of the National Society for the Study of Education: Part I: Second Report of the Society's Committee on New Materials of Instruction*, edited by G. M. Whipple, 163–164. Bloomington, IN: Public School Publishing Company, 1921.

Croft, Terrell. *Wiring of Finished Buildings: A Practical Treatise Dealing with the Commercial and the Technical Phases of the Subject, for the Central-*

Station Man, Electrical Contractor and Wireman. New York: McGraw-Hill, 1915.

Cutler-Hammer Manufacturing Company. "Easy—One Hand." *National Electrical Contractor* 16 (1916): 98.

Cutler-Hammer Manufacturing Company. "Untitled." *Motor Age* 25 (1914): 4.

Daily Inter-Ocean. "Another Button Scheme." 1892.

Dana, John Cotton. "Anticipations, or What We May Expect in Libraries." *Public Libraries* 12, no. 10 (1907): 381–383.

Dant, Tim. "Morality and Materiality." Paper presented at Design and Consumption: Ideas at the Interface. Durham University, London, January 12–13, 2005.

Darlington, Jennie. "Science for Children." *Pennsylvania School Journal* 38 (1889): 170.

Das Telephon. "American Notes." *Telegraphic Journal and Electrical Review* 24, no. 582 (1889): 71–72.

Dawson, Benjamin E. "Orificial Surgery as a Prophylactic Measure." *Journal of the American Association of Orificial Surgeons* 2, no. 1 (1914): 36–40.

Dawson, B. E. "My Front Door Key." *Medical Brief* 35 (1907): 901–902.

Day, Charles S. *A New Illustrated and Descriptive Catalogue, Illustrating and Describing Many Attractive Novelties, Electrical Goods, Useful Articles, Fancy Goods, Fine Jewelry, &C.* New Market, NJ: Charles S. Day, ca. 1890. Collections of the Winterthur Library, Winterthur, DE.

de Certeau, Michel, Luce Giard, and Pierre Mayol. *The Practice of Everyday Life: Living and Cooking.* Vol. 2. Minneapolis: University of Minnesota Press, 1998.

de la Peña, Carolyn Thomas. *The Body Electric: How Strange Machines Built the Modern American.* New York: New York University Press, 2003.

De Soto. "The Car That Has All Its Buttons." *Life* 45, no. 25 (1959): 11.

Dean Electric Co. "Tuto Horn." *Literary Digest* 44 (1912): 1064.

Dean, Mary, on behalf of William Dean (deceased). "Push Button Switch." November 24, 1925.

Dew-Smith, Alice. *The Diary of a Dreamer*. New York: G. P. Putnam's Sons, 1900.

Dickens, Charles. "What There Is in a Button." In *Household Words*. Vol. 5. Edited by Charles Dickens. London: 1852.

Dickson, J. L. "How to Make an Aluminum Push-Button." *Science and Industry* 7 (1907): 42.

Diemer, Hugo. *Factory Organization and Administration*. New York: McGraw-Hill, 1910.

Dodge, Martin, and Rob Kitchin. "Towards Touch-Free Spaces: Sensors, Software and the Automatic Production of Shared Public Toilet." In *Touching Space, Placing Touch*, edited by Mark Paterson and M. Dodge. Burlington, VT: Ashgate, 2012.

Drysdale, William. *Helps for Ambitious Boys*. New York: Thomas Y. Crowell Company, 1899.

Dunn, E. L. "Electric Dumb-Waiter Machines and Systems." *Electrical World* 50, no. 5 (1907): 223.

Duval, Wm. C., Jr. "Concrete Expression." *Bulletin of Photography* 14, no. 345 (1914): 331–334.

Eastman Kodak Company. "The Witchery of Kodak." In *Ellis Collection of Kodakiana*, edited by P. Wayne. Durham, NC: Duke University, 1911.

Eastman Kodak Company. "Take a Kodak with You." In *Ellis Collection of Kodakiana*, edited by P. Wayne. Durham, NC: Duke University, 1908.

Eastman Kodak Company. "Told by the Kodak." In *Ellis Collection of Kodakiana*, edited by P. Wayne. Durham, NC: Duke University, 1907.

Eberlein, Harold Donaldson. "The Revival of Bell Pulls." *House Beautiful* 39, no. 2 (1916): 38–39.

Edison Company. "Philosophic Bits." In *Question Box and Wrinkles*, edited by Homer E. Niesz and H. C. Abell, 31. Vol. 2 of the *National Electric Light Association Twenty-Eighth Convention*. New York: James Kempster Printing Company, 1905.

Edison Electric Illuminating Company. *Solid Comfort, or the Matchless Man: A Modern Realistic Story in Two Parts*. New York: Edison Electric Illuminating Company, 1903. Warshaw Collection of Business Americana. Archives Center, National Museum of American History, Smithsonian Institution, Washington, DC.

Edison Electric Illuminating Company. "The Edison Man." 1906. Warshaw Collection of Business Americana. Archives Center, National Museum of American History, Smithsonian Institution, Washington, DC.

Edison Electric Illuminating Company. *House Wiring: Requirements and Recommendations of the Edison Electric Illuminating Company of New York*. New York: Edison Electric Illuminating Company, 1892.

Edwards, Nina. *On the Button: The Significance of an Ordinary Item*. London: I. B. Tauris, 2011.

Edwinson, George. "Electric Bells." *Amateur Work, Illustrated* 1 (1883): 517–521.

"Eighteenth Precinct Police Station." *Documents of the Senate of the State of New York* 31 (1917): 153.

Electrical Merchandising, ed. *How to Sell Electrical Labor-Saving Appliances: 119 Tested Plans for the Electric Store*. New York: McGraw-Hill, 1918.

Electrical Record. *Special Devices for Electrical Uses*. New York: Gage Publishing, 1917.

Electrical Specialty Co. *Manufacturers and Owners of Woltmann & Trigg's Celebrated "Standard" Single Push Button Flush Switch*. Denver, CO: The Electrical Specialty Co., 1893. Collections of the Winterthur Library, Winterthur, DE.

The Electric Construction and Supply Co. "Private Residences and Apartments." n.d. Warshaw Collection of Business Americana. Archives Center, National Museum of American History, Smithsonian Institution, Washington, DC.

Electric Signal Clock Co. of Harrisburg and Waynesboro, PA. Illustrated General Catalogue. 1891. Warshaw Collection of Business Americana. Archives Center, National Museum of American History, Smithsonian Institution, Washington, DC.

Electric Vehicle Company. "Columbia Electric Carriage." *Town and Country*, September 15, 1906: 29.

"Electrified Gas Cars." *Electric Vehicles* 11, no. 6 (1917): 168.

Ellis, Havelock. *Studies in the Psychology of Sex: Erotic Symbolism, the Mechanism of Detumescence, the Psychic State in Pregnancy*. Philadelphia: F. A. Davis Company, 1914.

Emery, David. "Is President Trump's Nuclear Button Bigger Than Kim Jong-un's?" *Snopes*, January 3, 2018. https://www.snopes.com/2018/01/03/nuclear-button-trump-kim/.

Essig, Mark. *Edison and the Electric Chair: A Story of Light and Death*. New York: Walker, 2003.

Estes, Adam Clark. "Steve Jobs Didn't Want Apple Devices to Have an Off Switch." *Wire*, October 24, 2011. http://www.thewire.com/technology/2011/10/isaacson-jobs-60-minutes-interview/44028.

Evening Bulletin. "Successor of the Chestnut Bell." 1886: 4.

Eveready Corporation. "Don't Grope in the Dark." 1915. Warshaw Collection of Business Americana. Archives Center, National Museum of American History, Smithsonian Institution, Washington, DC.

The Factory Management Series: Machinery and Equipment. Chicago: A. W. Shaw Company, 1915.

Farhi, Paul. "So, You Had Questions about Matt Lauer's Desk?" *Washington Post*, December 1, 2017. https://www.washingtonpost.com/lifestyle/style/so-you-had-questions-about-that-button-on-matt-lauers-desk/2017/12/01/48b1f7c2-d6bd-11e7-a986-d0a9770d9a3e_story.html.

Fellman, Megan. "Bringing Texture to Your Flat Touchscreen." *North-western University News*, February 9, 2015. http://www.northwestern.edu/newscenter/stories/2015/02/bringing-texture-to-your-flat-touchscreen.html. Accessed March 31, 2016.

Findlay, Alexander. *Chemistry in the Service of Man*. London: Longmans, Green and Co., 1920.

Finley, Klint. "Press Amazon's IRL Button to Call a Cab, Brew Coffee, or Whatever." *Wired*, May 13, 2016. http://www.wired.com/2016/05/aws-iot-dash/.

Fischer, Claude S. *America Calling: A Social History of the Telephone to 1940*. Berkeley: University of California Press, 1992.

Floyd, Janet. *Domestic Space: Reading the Nineteenth-Century Interior*. Manchester, UK: Manchester University Press, 1999.

Flusser, Vilém. *Shape of Things: A Philosophy of Design*. London: Reaktion Books, 1999.

Forbes MFG. "Push the Button." *Hopkinsville Kentuckian*, February 18, 1913: 8.

Forster, Edward Morgan. "The Machine Stops." In *Selected Stories*, edited by David Leavitt and Mark Mitchell, 91–123. New York: Penguin Books, 2001.

Foucault, Michel. *Essential Works of Foucault 1954–1984: Power*. Vol. 3. London: Penguin Books, 2000.

Foucault, Michel. *Power/Knowledge: Selected Interviews and Other Writings 1972–1977*. New York: Pantheon Books, 1980.

Frances-Rand Company. "Safety for Fords." *Automobile Trade Journal* 22 (1909): 313.

Frank H. Stewart Electric Company. *Our New Home and Old Times*. Philadelphia: Frank H. Stewart Electric Company, 1913. Collections of the Huntington Library, San Marino, CA.

Frederick, Christine. *The New Housekeeping: Efficiency Studies in Home Management*. Garden City, NJ: Doubleday, Page & Company, 1913.

Galveston Daily News. "What One Electric Bell Did." November 4, 1889: 6.

Garfield, E. "The Tyranny of the Horn—Automobile, That Is." *Essays of an Information Scientist* 6, no. 11 (1983): 216–222.

Gas World. "'Telephos' Redivivus. Mechanical Difficulties Surmounted." 1908: 105.

Gates, S. J. "Electricity in the Home." *Wisconsin Engineer* 20, no. 7 (1916): 301–309.

Gelber, Steven M. *Hobbies: Leisure and the Culture of Work in America.* New York: Columbia University Press, 1999.

Gelber, Steven M. "Do-It-Yourself: Constructing, Repairing and Maintaining Domestic Masculinity." *American Quarterly* 49, no. 1 (1997): 66–112.

Gellner, Ernest. *Legitimation of Belief.* Cambridge: Cambridge University Press, 1975.

General Automobile Supply. *Catalogue.* 1909. Warshaw Collection of Business Americana. Archives Center, National Museum of American History, Smithsonian Institution, Washington, DC.

General Electric. "Christmas Electric Lighting." ca. 1910. Warshaw Collection of Business Americana. Archives Center, National Museum of American History, Smithsonian Institution, Washington, DC.

General Electric. *Electricity on the Farm.* 1913. Warshaw Collection of Business Americana. Archives Center, National Museum of American History, Smithsonian Institution, Washington, DC.

General Electric. "G.E. Removable Mechanism Switch." In *Juice: Live Information About Electric Goods.* Boston: Pettingell-Andrews, 1912. Collections of the Huntington Library, San Marino, CA.

General Electric. *G.E. Specialties.* 1906. Warshaw Collection of Business Americana. Archives Center, National Museum of American History, Smithsonian Institution, Washington, DC.

General Electric. *The Home of a Hundred Comforts*. Bridgeport, CT: General Electric, ca. 1920. Collections of the Winterthur Library, Winterthur, DE.

General Electric. "The Mazda Lamp in the Home." 1910.

Gerhard, William Paul. "The Essential Conditions of Safety in Theatres.— IV." *American Architect and Building News* 45, no. 969 (1894): 25–26.

Gerhard, William Paul. *Theater Fires and Panics*. New York: John Wiley & Sons, 1897.

Gerhard, William Paul. *Gas-Lighting and Gas-Fitting, Including Specifications and Rules for Gas Piping, Notes on the Advantages of Gas for Cooking and Heating, and Useful Hints to Gas Consumers*. New York: D. Van Nostrand Company, 1894.

Geyer, Jack. "The Pillow Pilots of the Push-Button Age." *Los Angeles Times*, September 24, 1952: A5.

Gibson, Charles Robert. *Electricity of To-Day: Its Work & Mysteries Described in Non-Technical Language*. London: Seely, Service & Co. Ltd., 1912.

Giddens, Anthony. *A Contemporary Critique of Historical Materialism: Power, Property and the State*. Berkeley: University of California Press, 1985.

Giedion, Sigfried. *Mechanization Takes Command: A Contribution to Anonymous History*. Oxford: Oxford University Press, 1948.

Gillette, Harry Orrin. "A Point of View in the Teaching of Electricity in the University Elementary School." *Elementary School Journal* 6 (1906): 306–309.

Gitelman, Lisa. *Scripts, Grooves, and Writing Machines: Representing Technology in the Edison Era*. Stanford, CA: Stanford University Press, 1999.

Gooday, Graeme. *Domesticating Electricity: Technology, Uncertainty and Gender, 1880–1914*. London: Pickering & Chatto, 2008.

Gordon, J. E. H. *Decorative Electricity*. London: Sampson Low, Marston, Searle, & Rivington, 1891. Collections of the Bakken Museum, Minneapolis, MN.

Granovetter, Mark, and Patrick McGuire. "The Making of an Industry: Electricity in the United States." In *The Law of Markets*, edited by Michael Callon, 147–173. Oxford: Blackwell, 1998.

Granville, Augustus Bozzi. *A Letter to the Right Hon. F. Robinson, President of the Board of Trade, and Treasurer of the Navy, on the Plague and Contagion with Reference to the Quarantine Laws*. London: Burgess and Hill, 1819.

Gray, Elisha. *Nature's Miracles: Familiar Talks on Science*. Vol. 2. New York: Baker & Taylor, 1900.

Greeley, E. S. "Electricity Applied to Household Affairs." *Independent: A Weekly Journal of Free Opinion* 45 (1893): 7–8.

Greenwald, Will. "Hands on with the Oculus Rift and Oculus Touch Controllers." *PC Magazine*, January 7, 2016. https://www.pcmag.com/article2/0,2817,2497577,00.asp.

Gregory, John Milton. *A New Political Economy*. Cincinnati, OH: Bragg & Co., 1882.

Grint, Keith. "Introduction." In *The Sociology of Work*. Cambridge: Polity Press, 2005.

Grint, Keith, and Steve Woolgar. "Computers, Guns, and Roses: What's Social about Being Shot?" *Science, Technology & Human Values* 17, no. 3 (1992): 366–380.

Grover Brothers. *Popular Electric Lighting, a Few Practical Hints to Present and Intending Users of ... Electrical Energy for Illuminating Purposes and for Power, and Showing Some of the Latest Methods of ... Heating and Cooking by Electricity*. Newark, NJ: Farrand, 1898.

Guy, George Heli. "Electricity in the Household." *Chatauquan* 26, no. 1 (1897): 50.

Hale, R. S. "Unutilized Comforts of Electricity." *American Gas Light Journal* 73, no. 14 (1900): 530.

Hall, Wendy. "Ending the Tyranny of the Button." *IEEE MultiMedia* 1, no. 1 (1994): 60–68.

Ham, Charles H. *Mind and Hand: Manual Training the Chief Factor in Education.* New York: American Book Company, 1900.

Hammer, William J. *Electric Diablerie: Being a Veracious Account of an Electrical Dinner Tendered in 1884 by William J. Hammer, Consulting Electrical Engineer to the "Society of Seventy-Seven" of the N.P.H.S. of Newark, N.J., in the First Electrical House Ever Established Anywhere in the World.* New York: William J. Hammer, 1885. Warshaw Collection of Business Americana. Archives Center, National Museum of American History, Smithsonian Institution, Washington, DC.

Hammond, Robert. *The Electric Light in Our Homes.* London: Frederick Warne and Co., 1884.

"Hand-Picked Members." *Mission Studies* 2 (1915): 6–8.

Hardyment, Christina. *From Mangle to Microwave: The Mechanization of Household Work.* Cambridge: Polity Press, 1988.

Hargardon, Andrew B., and Yellowlees Douglas. "When Innovations Meet Institutions: Edison and the Design of the Electric Light." *Administrative Science Quarterly* 46, no. 3 (2001): 476–501.

Harper's Weekly. "To Prevent Panic in Theatres." April 22, 1882: 243.

Harris, Neil, Wim de Wit, James Burkhart Gilbert, and Robert W. Rydell, and the Chicago Historical Society. *Grand Illusions: Chicago's World's Fair of 1893.* Chicago: Chicago Historical Society, 1993.

Hart, Hastings Hornell. *Cottage and Congregate Institutions for Children.* New York: Charities Publication Committee, 1910.

Hart and Hegeman Manufacturing Company. "The Hart & Hegeman Mfg. Co. Electric Switches and Wiring Devices." *Electrical Review* 38, no. 2 (1901): 35.

Hasluck, Paul N. *Electric Bells: How to Make and Fit Them; Including Batteries, Indicators, Pushes, and Switches*. Philadelphia: David McKay, 1914.

Hayes, C. M. "Minutes of Convention of Photographers' Association of America, Held July 12–17, 1897, at Celeron-on-Chautauqua, Chautauqua County, NY." *Anthony's Photographic Bulletin* 28, no. 7 (1897): 234–261

Hayles, N. Katherine. *How We Became Posthuman: Virtual Bodies in Cybernetics, Literature, and Informatics*. Chicago: University of Chicago Press, 1999.

Heckman, Davin. *A Small World: Smart Houses and the Dream of the Perfect Day*. Durham, NC: Duke University Press, 2008.

Herrick, Christine Terhune. *The Expert Maid-Servant*. New York: Harper & Brothers, 1904. Collections of the Winterthur Library, Winterthur, DE.

Herzog Telesme. "Untitled." n.d. Warshaw Collection of Business Americana. Archives Center, National Museum of American History, Smithsonian Institution, Washington, DC.

Higginbotham, Helena. "The Electric Bell a Woman's Charge." *Good Housekeeping* 40 (1905): 642–644.

Hill, M. J. "The Clitoris." *Homoeopathic Journal of Obstetrics, Gynaecology and Pediatrics* 18 (1896): 582–583.

Hill, W. R. "Door Hardware for the Modern Home." *Building Age* 41, no. 8 (1919): 253–254.

Hindman, Hugh D. *The World of Child Labor: An Historical and Regional Survey*. Armonk, NY: M. E. Sharpe, 2009.

Hine, Thomas. *Populuxe*. New York: Knopf, 1986.

Hoadley, G. L. "Home Electrics." *New Science and Invention in Pictures* 9, no. 1 (1921): 67.

Hoar, Jennifer. "Apple's Steve Jobs Hates Buttons." *CBS News*, July 25, 2007. http://www.cbsnews.com/news/apples-steve-jobs-hates-buttons/.

Holden, Edward S. *Real Things in Nature: A Reading Book of Science for American Boys and Girls.* New York: Macmillan, 1903.

Holden, Edward S. *The Sciences: A Reading Book for Children.* Boston: Ginn & Company, 1903.

Hool, George A. *Reinforced Concrete Construction.* New York: McGraw-Hill, 1927.

Hopkins, Emma Curtis. *Spiritual Law in the Natural World.* Chicago: Purdy Publishing, 1894.

Hopkins, George M. *Home Mechanics for Amateurs.* New York: Munn & Co., 1903.

Hopkinson, William B. "Electric Fire-Alarm." Patent No. 769,824. September 13, 1904.

Hopley, Frank Dorrance. "The 'Push-the-Button Man.'" *Chicago Daily Tribune,* June 8, 1924: B6.

Hospitalier, Eduoard, and C. J. Wharton. *Domestic Electricity for Amateurs.* London: E. & F. N. Spon, 1889.

Hough, Lynn Harold. "The Age of Machinery." *Michigan Technic* 33 (1920): 179–182.

House & Garden. "The House Telephone." 1911: 269.

Houston, Edwin J. *Electricity in Every-Day Life.* New York: P. F. Collier & Son, 1905.

Houston, Edwin J. *A Dictionary of Electrical Worlds, Terms and Phrases.* New York: W. J. Johnston, 1892.

Hubert, Philip Gengembre. "A Letter to the Rising Generation." *Atlantic Monthly* 107 (1911): 147.

Hughes, Alice. "The Woman's Slant." *Life* 99 (1932): 42.

Hughes Electric Heating Co. "The Hughes Electric Cook Stove." 1910. Warshaw Collection of Business Americana. Archives Center, National Museum of American History, Smithsonian Institution, Washington, DC.

358   References

Hughes, Thomas Parke. *Networks of Power: Electrification in Western Society, 1880–1930*. Baltimore: Johns Hopkins University Press, 1983.

Huhtamo, Errki. "Push the Button, Kinoautomat Will Do the Rest! Media-Archaeological Reflections on Audience Interactivity." In *Expanding Practices in Audiovisual Narrative*, edited by Raivo Kelomees and Chris Hales. Newcastle upon Tyne, UK: Cambridge Scholars Publishing, 2014.

Huhtamo, Erkki. "From Kaleidoscomaniac to Cybernerd: Towards an Archeology of the Media." http://web.stanford.edu/class/history34q/readings/MediaArchaeology/HuhtamoArchaeologyOfMedia.html.

Humfreville, James Lee. *Twenty Years among Our Savage Indians: A Record of Personal Experiences, Observations, and Adventures among the Indians of the Wild West*. Hartford, CT: The Hartford Publishing Company, 1897.

Hunt, Clara Whitehill. *Library Work with Children*. New York: American Library Association, 1924.

Hunter, Rudolph M. "Push Button." Patent No. 510. December 12, 1893: 540.

Hussobee, Martin. "Electric Light Companies Appeal to Public in Big Campaign." *Printers' Ink* 114 (1921): 10–17.

Hyde, Henry M. "The Automatic General Manager." *Business, a Magazine for Office, Store, and Factory* 19 (1906): 31–36.

Iles, George. "Electricity as a Domestic." *Everybody's Magazine* (1901): 344.

Independent. "Touching the Button." May 11, 1893: 10.

Inglesby, Pamela. "Button-Pressers Versus Picture-Makers: The Social Reconstruction of Amateur Photography in the Late 19th Century U.S." *Visual Sociology* 5, no. 1 (1990): 18–25.

Jackson, P., and J. Lethem, eds. *The Exegesis of Philip K. Dick*. Boston: Houghton Mifflin Harcourt, 2011.

James, Dorothy. "The Building of a House." *Christian Union* 40, no. 26 (1889): 845.

Janlert, Lars-Erik. "The Ubiquitous Button." *Interaction* 21, no. 3 (2014): 26.

J. Elliott Shaw Company. *Household and Experimental Electrical Supplies and Novelties, Catalog No. 15.* 1903. Warshaw Collection of Business Americana. Archives Center, National Museum of American History, Smithsonian Institution, Washington, DC.

Jenkins, Reese V. "Technology and the Market: George Eastman and the Origins of Mass Amateur Photography." *Technology and Culture* 16, no. 1 (1975): 1–9.

Jepson, George. "Push-Button." Patent No. 574,247. December 29, 1896.

Jerrold, Walter. *Electricians and Their Marvels.* New York: Fleming H. Revell Company, 1893. Collections of the Huntington Library, San Marino, CA.

Jewett, Frances Gulick. *Control of Body and Mind.* Boston: Ginn & Company, 1908.

J. H. Bunnell & Company. *Illustrated Catalogue and Price List of Telegraphic, Electrical & Telephone Supplies No. 9.* 1888. Warshaw Collection of Business Americana. Archives Center, National Museum of American History, Smithsonian Institution, Washington, DC.

J. J. Duck Company. *Anything Electrical, Catalog No. 6.* Toledo, OH: J. J. Duck Company, 1912. Collections of the Bakken Museum, Minneapolis, MN.

Johnson, Joseph French. "Advertising as an Economic Force." *Judicious Advertising* 18 (1920): 27–31.

Johnson, Louis F. "Electric Push-Button." Patent No. 631,892. 1899.

Johnston, Patricia A. *Real Fantasies: Edward Steichen's Advertising Photography.* Berkeley: University of California Press, 1997.

Jones, W. Clyde. "The $25 Prize Essay. How Can the Department of Electricity at the World's Columbian Exposition Best Serve the Electrical Interests?" *World's Fair Electrical Engineering, an Illustrated Monthly Magazine* 1, no. 3 (1893): 129–135. Warshaw Collection of Business

Americana. Archives Center, National Museum of American History, Smithsonian Institution, Washington, DC.

Kasson, John F. *Amusing the Million: Coney Island at the Turn of the Century*. New York: Hill & Wang, 1978.

Kelsey, J. C. "Some Steps in the Evolution of Circuit Design—Article X—Introductory to Common Battery, or Central Energy Systems." *American Telephone Journal* 7 (1903): 131–132.

Kennedy, Rankin. *The Book of Electrical Installations*. Vol. 2. London: Caxton Publishing, 1902.

Kennelly, Arthur E. "The Evolution of Electric and Magnetic Physics." In *Evolution in Science, Philosophy, and Art*, edited by Brooklyn Ethical Association. New York: D. Appleton and Company, 1891.

Kent, Ernest B. "The Elementary School and Industrial Occupations." *Elementary School Teacher* 9, no. 4 (1908): 178–185.

Kern, Stephen. *The Culture of Time and Space, 1880–1918: With a New Preface*. Cambridge, MA: Harvard University Press, 2003.

Kidder, Frank Eugene. *The Architect's and Builder's Pocket-Book: A Handbook for Architects*. New York: John Wiley & Sons, 1908.

Klare, Michael T. "Whose Finger on the Nuclear Button?" *Common Dreams*, November 6, 2016. https://www.commondreams.org/views/2016/11/06/whose-finger-nuclear-button/.

Knapp, Philip Coombs. *Accidents from the Electric Current: A Contribution to the Study of the Action of Currents of High Potential Upon the Human Organism*. Boston: Damrell & Upham, 1890.

Knibbs, Kate. "People Got So into This Strange Internet Button They Made Up Religions." *Gizmodo*, May 12, 2015. https://gizmodo.com/people-got-so-into-this-strange-internet-button-they-ma-1700779699.

Krohn, William Otterbein. *Practical Lessons in Psychology*. Chicago: The Werner Company, 1895.

Kuehn, V. A. *Bells and Annunciators*. Architect and Engineer 44, no. 1 (1916): 117–118.

Kuttner, Alfred. "Freud's Contribution to Ethics." *New Republic* 5 (1915): 101–103.

Latour, Bruno. "Mixing Humans with Non-Humans: Sociology of a Door-Closer." In *Ecologies of Knowledge—Work and Politics in Science and Technology*, edited by Susan Leigh Star, 298–310. Albany, NY: SUNY Press, 1995.

Latour, Bruno. "The Force and Reason of Experiment." In *Experimental Inquiries, Historical, Philosophical and Social Studies of Experimentation in Science*, edited by Homer Le Grand, 48–79. Dordrecht, the Netherlands: Kluwer Academic Publishers, 1990.

Latour, Bruno. *Science in Action: How to Follow Scientists and Engineers through Society*. Cambridge, MA: Harvard University Press, 1988.

Latour, Bruno. "Technology Is Society Made Durable." In *A Sociology of Monsters? Essays on Power, Technology and Domination, Sociological Review Monograph*, edited by John Law, 103–131. London: Routledge, 1991.

Lavrinc, Damon. "Apple's New Trackpad Might Be the Solution to the Touchscreen Problem." *Jalopnik*, March 24, 2015. http://jalopnik.com/apple-s-new-trackpad-might-be-the-solution-to-the-touch-1693383080. Accessed January 15, 2016.

Lee, Gerald Stanley. *The Voice of the Machines: An Introduction to the Twentieth Century*. Northampton, MA: Mount Tom Press, 1901. Collections of the Huntington Library, San Marino, CA.

Leech, Alfred. "Your New Home of 1980 May Operate by Thumb." *Washington Post*, December 8, 1956: C4.

Leffingwell, W. H. "Increase Human Efficiency." *Natural Gas and Gasoline Journal* 13, no. 3 (1919): 190.

Lemmon, George T. *The Eternal Building or the Making of Manhood*. New York: Eaton & Mains, 1899.

Lent, Frank Townsend. *Sound Sense in Suburban Architecture: Containing Hints, Suggestions, and Bits of Practical Information for the Building of Inexpensive Country Houses*. New York: W. T. Comstock, 1895.

Leonard, H. Ward. *Electricity in a Modern Residence*. New York: H. Ward Leonard & Co., 1892.

Leslie, Eliza. *The House Book, or, a Manual of Domestic Economy: For Town and Country*. Philadelphia: Carey & Hart, 1845.

Leslie, Eliza. *Miss Leslie's Behavior Book: A Guide and Manual for Ladies as Regards Their Conversation; Manners; Dress; Introductions; Entree to Society; Shopping; Conduct in the Street; at Places of Amusement. In Traveling; at the Table, Either at Home, in Company, or at Hotels; Deportment in Gentlemen's Society; Lips; Complexion; Teeth; Hands, the Hair; Etc., Etc.* Philadelphia: T. B. Peterson and Brothers, 1839.

Lockwood, Mary Smith. *Historic Homes in Washington: Its Noted Men and Women*. New York: Belford Company, 1889.

Lockwood, Thos. D. "Electrical Killing." *Electrical Engineer* 7, no. 75 (1888): 89–90.

Logan Republican. "Push Buttons." September 12, 1914: 6.

Logitech. "Introducing the Most Agile Mouse Ever to Set Foot on a Desktop." *Byte* (1998): 1.

Los Angeles Times. "The Big Blast." October 11, 1885: 1.

Los Angeles Times. "A Button for the Baby." September 18, 1895: 1.

Los Angeles Times. "Electric Safety Cabinet." May 31, 1897: 9.

Los Angeles Times. "Electric Slave to Serve Homes." May 18, 1932.

Los Angeles Times. "Even Crossing Street Can Be Big Problem in Push-Button World." December 10, 1961: WS5.

Los Angeles Times. "The Girlless Telephone." April 12, 1903: D11.

Los Angeles Times. "Her Dainty Touch." July 24, 1895: 1.

Los Angeles Times. "His Own Elevator Operator." August 21, 1898: B3.

Los Angeles Times. "The Kitty's Secrets Out." June 27, 1901: 9.

Los Angeles Times. "Press the Button." September 29, 1922: II4.

Los Angeles Times. "Self-Made Men." March 31, 1925: A4.

Los Angeles Times. "Signal Lamps Call Police." August 12, 1904: 6.

Loughead, Flora Haines. "The House on the Hill: The Doorbell Tells the Story." *Overland Monthly* 15, no. 85 (1890): 64–72.

Lundquist, Ruben Alvin. *Electrical Goods in British India and Ceylon. Special Agent Series—No. 213.* Washington, DC: Government Printing Office, 1922.

Lundquist, Ruben Alvin. "Electrical Goods in British South Africa." In *Department of Commerce. Special Agents Series—No. 205.* Washington, DC: Government Printing Office, 1920.

Lundquist, Ruben Alvin. *Electrical Goods in Australia. Special Agents Series—No. 155.* Washington, DC: Government Printing Office, 1918.

Macdonald, Fiona. *Victorian Servants, a Very Peculiar History.* Brighton, UK: Salariya, 2011.

MacKenna, Robert William. *The Adventure of Death.* New York: G. P. Putnam's Sons, 1917.

Mailloux, Cyprien. "Push-Button." Patent No. 575,523. January 19, 1897.

Maine Farmer. "Electric Bells." June 29, 1880.

Maines, Rachel. *The Technology of Orgasm: "Hysteria," the Vibrator, and Women's Sexual Satisfaction.* Johns Hopkins Studies in the History of Technology, No. 24. Baltimore: Johns Hopkins University Press, 1999.

Manhattan Electrical Supply. *Something Electrical for Everybody, Catalogue Twenty-Six.* New York: Manhattan Electrical Supply Co., ca. 1915

Manhattan Electrical Supply. "Electrical, Bicycle, and Photographic Supplies." n.d. Warshaw Collection of Business Americana. Archives Center, National Museum of American History, Smithsonian Institution, Washington, DC.

Manhattan Electrical Supply. "The Philosophy and Practice of Morse Telegraph; Also Illustrations, Descriptions and Price List of Something

Electrical for Everybody." n.d. Warshaw Collection of Business Americana. Archives Center, National Museum of American History, Smithsonian Institution, Washington, DC.

Manovich, Lev. *The Language of New Media*. Cambridge, MA: MIT Press, 2001.

Manson, Ray H. "Push-Button Switch." Patent No. 839,215. March 4, 1919.

Marillier, Harry C. "The Automobile: A Forecast." *Eclectic Magazine of Foreign Literature, Science, and Art* 62, no. 6 (1895): 774–780.

Marks, Edward Charles Robert. *Notes on the Construction of Cranes and Lifting Machinery*. Manchester, UK: Technical Publishing Co., 1904.

Marshall, J. E. "School Hygiene and Sanitation." *Proceedings of the Fifty-Fourth Annual Session of the Iowa State Teachers Association* (1909): 107–110.

Martin, Robert E. "Each of Us Has 40 Slaves." *Popular Science Monthly* 111, no. 2 (1927): 26, 131.

Martin, T. C. "The Work and Responsibilities of the 'Local Electrician.'" *Electrical Engineer* 10, no. 133 (1890): 568.

Martin, T. Commerford, and Stephen Leidy Coles, eds. *The Story of Electricity*. Vol. 1. New York: The Story of Electricity Company, 1919.

Martineau, C. A. "Royal Victoria Hall." *Knowledge* 11 (1888): 136–137.

Marvin, Carolyn. *When Old Technologies Were New: Thinking About Electric Communication in the Late Nineteenth Century*. New York: Oxford University Press, 1988.

Marx, Karl, and Friedrich Engels. *Manifesto of the Communist Party*. London: Reeves, 1888.

Masten, F. J. "Two Speeds for an Engine." *Wood-worker* 12 (1893): 15.

Matheson, Richard. "Button, Button." In *Collected Stories*. Vol. 3. Edited by Stanley Wiater. Colorado Springs, CO: Gauntlet Press, 2005.

Mattern, W. M. P. "To Boost the Beginner." *American Photography* 16 (1922): 152.

Maver Jr., William. *William Maver's Wireless Telegraphy: Theory and Practice.* New York: Maver Publishing Company, 1904.

Maycock, William Perren. *Small Switches, Etc., and Their Circuits.* London: S. Rentell & Co., 1911.

McBride's Magazine. "Walnuts and Wine." 1907: 885.

McCormick, W. H. *Electricity.* London: Frederick A. Stokes, 1915.

McGraw-Hill Book Company. *Wiring Diagrams of Electrical Apparatus and Installations.* New York: McGraw-Hill, 1913.

McIntire, A. H. "The Pushbutton Habit." *Electricity Journal* 14, no. 2 (1917): 41–42.

McLaughlin, J. C. "Electrical Call Systems for Hotel." Patent No. 335,604. February 9, 1886.

McLaughlin, James F. "Push-Button." Patent No. 521,808. June 26, 1894.

McMeal, Harry B. "Push Button Telephones." *Telephony* 3–4 (1902): 329–330.

McNeff, L. Dow. "Electricity as a Subject for Study in Elementary Schools. Part I." *Elementary School Teacher* 8, no. 5 (1908): 271–276.

McQuiston, J. C. "Reason for Change in Westinghouse Policy." *Printers' Ink* 84 (1913): 78.

Medak, Peter, dir. *The Twilight Zone.* Season 1, episode 50, "Button, Button." Aired March 7, 1985, on CBS.

Meekins, Lynn W. "World Markets for American Manufacturers." *Scientific American* 120, no. 1 (1919): 12.

Miller, Francis Trevelyan. *Wonder Stories.* New York: The Christian Herald, 1913.

Miller, Morris G. "Switch Arm for Use on Basement-Light Circuit." *Popular Mechanics* 23, no. 5 (1915): 764.

Milliken, George F. "Fire-Alarm Box." Patent No. 412,971. 1889.

Moore, Francis Cruger. *How to Build a Home: Being Suggestions as to Safety from Fire, Safety to Health, Comfort, Convenience, Durability and Economy.* New York: Doubleday & McClure, 1897.

Moore's Burglar Alarm Manufacturing Company. "Moore's Burglar Alarm Manufacturing Co." n.d. Warshaw Collection of Business Americana. Archives Center, National Museum of American History, Smithsonian Institution, Washington, DC.

Morgan, J. H. "Child Labor: Its Relation to Future Generations." In *Second Annual Report of the Department of Labor of the State of New York*, 575–577. Albany, NY: The Argus Company, 1903.

Morris, Charles. *The Nation's Navy: Our Ships and Their Achievements.* Philadelphia: J. B. Lippincott Company, 1898.

Morris, Robert T. "Is Evolution Trying to Do Away with the Clitoris?" *American Journal of Obstetrics and Diseases of Women and Children* 26 (1892): 847–858.

Mosso, Angelo. *Fatigue.* Trans. M. Drummond and W. B. Drummond. New York: G. P. Putnam's Sons, 1904.

Motor Car Equipment Company. *Automobile Accessories, Tools & Hardware Shop Equipment.* 1916. Warshaw Collection of Business Americana. Archives Center, National Museum of American History, Smithsonian Institution, Washington, DC.

"Motor Driven Grinder Tool." *Motor Age* 38 (1920).

Mrs. Motherly. *The Servants' Behaviour Book: Or Hints on Manners and Dress for Maid Servants in Small Households.* London: Bell and Daldy, 1859.

Munro, John. *The Story of Electricity.* New York: D. Appleton and Company, 1905.

Murray, Joseph John. "A Physical Theory of Moral Freedom." *Journal of the Transactions of the Victoria Institute* 22 (1889): 229.

Napier, John Russell, and Russell Tuttle. *Hands*. Rev. ed. Princeton: Princeton University Press, 1993.

National Association of Builders. "The Builder's Exchange." *Building Age* 15 (1893): 113.

National Association of Retail Druggists. "And There Are Others." *N.A.R.D. Notes* 10, no. 8 (1910): 524.

National Electric Light Association. "A Good Way to Sell Fans." *N.E.L.A. Bulletin* 4 (1917): 603.

Nesbit, W. B. "Push the Button." *Florida Star*, January 6, 1905: 2.

Newman, Hugo. "Science Teaching in Elementary Schools." *Elementary School Teacher* 6, no. 4 (1905): 192–202.

New York Times. "At the Touch of a Button: How Warships Are Controlled." September 13, 1896: BR13.

New York Times. "Calling the Servants." April 7, 1929: SM10.

New York Times. "A Child Can Operate the Otis Elevator." May 29, 1898: IMS16.

New York Times. "Death Shadowing Them." July 6, 1891: 1.

New York Times. "Display Ad 45." January 2, 1910: AU12.

New York Times. "Dr. Herz's New Telephone: Some Further Account of the Push-Button Device." May 1, 1887.

New York Times. "Golden Telegraph Key for President's Use." May 15, 1927: XX16.

New York Times. "Herz's Telephone." May 21, 1887.

New York Times. "In the World of Electricity." September 10, 1895: 3.

New York Times. "The Pan-American Fair: Mr. McKinley and the Presidents of Other Republics Will Start the Machinery on May 1." March 15, 1901: 9.

New York Times. "Rats Derange Electric Bells." January 21, 1898: 12.

New York Times. "Ripples of Radio News Eddying in the Ether." April 1, 1928: 157.

New York Times. "Trolleys in Larchmont." December 3, 1900: 1.

New York Times. "What Happens When a Fire Alarm Box Is 'Pulled.'" June 2, 1901: SM14.

Niven, Frederick. *The Porcelain Lady.* London: M. Secker, 1913.

Nogochi, Yone. *The American Diary of a Japanese Girl.* New York: Frank A. Stokes, 1902.

Noll, Augustus. "Lighting Residences by Electricity." *Engineering Magazine* 10, no. 6 (1896): 1059–1063.

Norman, James William. *Contributions to Education.* New York: Columbia University Teachers College, 1922.

Norman, Lloyd. "If War Comes It'll Be Long, Brucker Says." *Chicago Daily Tribune,* February 2, 1956: 12.

Norris, John G. "Calamity Howlers and Pollyannas Both Wrong on 'Push-Button War.'" *Washington Post,* August 3, 1947: B11.

North American. "Mrs. Cleveland's Touch." August 24, 1886.

North, S. N. D. *The Man and the Machine: A Plea for Industrial Education; an Address at the Commencement of the Pennsylvania Museum and School of Industrial Art.* Boston: Press of Rockwell and Churchill, 1896. Collections of the Huntington Library, San Marino, CA.

Northend, Mary H. "Reviving the Bell Pull." *House & Garden* 37, no. 2 (1920): 44–60.

Nunnamaker, Albert John, and Charles Otto Dhonau. *Hygiene and Sanitary Science: A Practical Guide for Embalmers and Sanitarians.* Cincinnati, OH: Embalming Book Company, 1913.

Nye, David E. *Electrifying America: Social Meanings of a New Technology, 1880–1940.* Cambridge, MA: MIT Press, 1990.

Oehring, A. J. "Circuit-Closer." Patent No. 502,749. August 8, 1893.

Ohio Electric Works. "An Electric Light for the Necktie." 1897. Warshaw Collection of Business Americana. Archives Center, National Museum of American History, Smithsonian Institution, Washington, DC.

Ohio Electric Works. "The $3 Electric Light for Necktie or Coat." n.d. Warshaw Collection of Business Americana. Archives Center, National Museum of American History, Smithsonian Institution, Washington, DC.

Oldenziel, Ruth. *Making Technology Masculine: Men, Women, and Modern Machines in America, 1870–1945*. Amsterdam: Amsterdam University Press, 1999.

O. S. Platt Manufacturing Company. "The 'New England' Push-Button Switch." *Electrical World* 29 (1897): 695.

O'Shea, M. V., and John Harvey Kellogg, eds. *Making the Most of Life*. New York: Macmillan, 1915.

Osherson, Samuel, and Lorna AmaraSingham. "The Machine Metaphor in Medicine." In *Social Contexts of Health, Illness, and Patient Care*, edited by Elliot G. Mishler, 218–249. Cambridge: Cambridge University Press, 1981.

Otis Elevator Company. "87 Years of ... Vertical Transportation with Otis Elevators." 1939. Warshaw Collection of Business Americana. Archives Center, National Museum of American History, Smithsonian Institution, Washington, DC.

Otter, Chris. *The Victorian Eye: A Political History of Light and Vision in Britain, 1800–1910*. Chicago: University of Chicago Press, 2008.

Oxley, George F. "More Than Two Million People View First N.E.L.A. Film—New Picture Is Shown at Convention." *N.E.L.A. Bulletin* 9, no. 6 (1922): 357–358.

Oxley, George F. "Movie Program Making Progress." *N.E.L.A. Bulletin* 8, no. 11 (1921): 660–661.

Packer, Jeremy. *Mobility without Mayhem: Safety, Cars, and Citizenship.* Durham, NC: Duke University Press, 2008.

Palmer, George Herbert. *The Nature of Goodness.* Boston: Houghton Mifflin Company, 1903.

Palmer, Granville E. "Good Switches Versus Poor Switches." *Electrical World* 58 (1911): 591.

Parce, Lida. *Economic Determinism: Or, the Economic Interpretation of History.* Chicago: Charles H. Kerr & Company, 1913.

Parikka, Jussi. *What Is Media Archaeology?* Cambridge: Polity Press, 2013.

Parisi, David P. "Tactile Modernity: On the Rationalization of Touch in the Nineteenth Century." In *Media, Technology, and Literature in the Nineteenth Century*, edited by Colette Colligan and Margaret Linley, 189–214. Farnham, UK: Ashgate, 2011.

Parke, John Y. "Electric Push-Button." Patent No. 727996. May 12, 1903.

Parker, J. R. "Buttons, Simplicity, and Natural Interfaces." *Loading ...* 2, no. 2 (2008). http://journals.sfu.ca/loading/index.php/loading/article/view/33/.

Paterson, Mark. *The Senses of Touch: Haptics, Affects and Technologies.* Oxford: Berg, 2007.

Patrick and Carter Co. *Patrick & Carter's Illustrated Catalogue and Price List.* Philadelphia: Patrick and Carter Co., 1882. Warshaw Collection of Business Americana. Archives Center, National Museum of American History, Smithsonian Institution, Washington, DC.

Pattison, Mary. *Principles of Domestic Engineering; or the What, Why and How of a Home.* New York: The Trow Press, 1915.

Peabody, Henrietta C., ed. "Homemakers' Questions and Answers: A Ready Reference for Those Who Are Building." In *Remodeling, Furnishing, Decorating or Gardening.* Boston: Atlantic Monthly Press, 1918.

Pearson, C. Arthur. *Things All Scouts Should Know: A Collection of 313 Illustrated Paragraphs of Useful Information, Specially Selected for the Use of Boy Scouts.* London: Limited, 1910.

Pearson, Edmund Lester. "New Books and Old." *Weekly Review* 4, no. 97 (1921): 275.

Peiss, Kathy Lee. *Cheap Amusements: Working Women and Leisure in New York City, 1880 to 1920*. Philadelphia: Temple University Press, 1986.

Perkins, Frank C. "Modern German Police Call and Fire Alarm Systems." *Scientific American* 71, no. S1836 (1911): 148–150.

Peters, Benjamin. "Digital." In *Digital Keywords: A Vocabulary of Information Society and Culture*, edited by Benjamin Peters, 93–108. Princeton, NJ: Princeton University Press, 2016.

Peters, John Durham. *Speaking into the Air: A History of the Idea of Communication*. Chicago: University of Chicago Press, 1999.

Peters, Mark. *Japan Dreams: Notes from an Unreal Country*. Bloomington, IN: Booktango, 2013.

Phillips, Cabell. "Why We're Not Fighting with Push Buttons." *New York Times*, July 16, 1950: SM7.

Phillips, David Graham. "The Story of George Helm." *Heart International* 22 (1912): 51–62.

Phillips, Eliot. "Staples Easy Button Hacking." *Hackaday*, July 18, 2006. http://hackaday.com/2006/07/18/staples-easy-button-hacking/.

Phillips, Morris. *Abroad and at Home: Practical Hints for Tourists*. New York: Brentano's, 1891.

"Play-Production by Push-Button." *Literary Digest* 60 (1919): 22.

Plotnick, Rachel. "Predicting Push-Button Warfare: U.S. Print Media and Conflict from a Distance, 1945–2010." *Media Culture & Society* 34, no. 6 (2012): 655–672.

Plotnick, Rachel. "Touch of a Button: Long-Distance Transmission, Communication and Control at World's Fairs." *Critical Studies in Media Communication* 30, no. 1 (2012): 52–68.

Plumb, Glenn E. "Plumb Plan for Government Ownership and Democracy in Operation of the Railroads." In *Proceedings of Public Ownership*

Conference, 59–109. Chicago: Public Ownership League of America, 1919.

Pold, Soren. "Button." In *Software Studies: A Lexicon*, edited by Matthew Fuller. Cambridge, MA: MIT Press, 2008.

Pratt, Loring. "Untitled." *Union Electric Quarterly* 5, no. 9 (1917): 328.

Prawdzik, Ben. "Milgram Experiment, 50 Years On." *Yale Daily News*, September 28, 2011. http://yaledailynews.com/blog/2011/09/28/milgram -experiment-50-years-on/.

Preece, W. H. "Recent Wonders of Electricity." *Popular Science* 20, no. 46 (1882): 786–794.

Preece, William Henry. "On the Application of Electricity to Domestic Purposes." *Telegraphic Journal* 1, no. 16 (1864): 181.

Price, Will. "What Is the Arts and Crafts Movement?" *International Wood Worker* 14 (1904): 354–356.

Price, William L. "The Attitude of Manual Training to the Arts and Crafts." *Proceedings of the Eastern Manual Training Association* (1903): 14–20.

"Proceeding of the City Council." February 4, 1907: 3003.

"Pullman 'Little Six.'" *Motor World* 39 (1914): 63.

Punch. "The Micro-Telephone Push-Button." 1887.

"Push (v.)." In *Oxford English Dictionary*. Oxford: Oxford University Press, 2016.

"Push-the-Button Habit." *Motor Age* 24, no. 5 (1913): 24.

Puskar, Jason. "Pistolgraphs: Liberal Technoagency and the Nineteenth-Century Camera Gun." *Nineteenth-Century Contexts* 36, no. 5 (2014): 517–534.

"The Randall Push Button." *Electrical Review* 14, no. 1 (1889): 26.

Ravndal, G. B. *Turkish Markets for American Hardware. Special Consular Reports—No. 77*. Washington, DC: Government Printing Office, 1917.

Raykoff, Ivan. "Piano, Telegraph, Typewriter: Listening to the Language of Touch." In *Media, Technology, and Literature in the Nineteenth Century: Image, Sound, Touch,* edited by Colette Colligan and Margaret Linley, 159–188. Farnham, UK: Ashgate, 2011.

"Report of the Special Committee on Local Transportation to the City Council." December 11, 1901: 7.

Resnick, Louis. "Speeding up Production with Safety in a Small Plant." *National Safety News* 3–4 (1921): 3–6.

Rice, Stephen P. *Minding the Machine: Languages of Class in Early Industrial America.* Berkeley: University of California Press, 2004.

Riis, Jacob August. *How the Other Half Lives: Studies among the Tenements of New York.* New York: Charles Scribner's Sons, 1890.

Ritter von Nussbaum, J. N. "The Influence of Antiseptics Upon Legal Medicine: A Lecture Delivered in the Course of Clinical Surgery in the Munich Hospital, in the Winter Session 1897–80." *Edinburgh Medical Journal* 26, no. 11 (1881): 993–1003.

Robbins, Alexander H. "Reasonableness and Legal Right of Minimum Charge in Public Utilities." *Central Law Journal* 85, no. 6 (1916): 95–98.

Roberts, Lissa. "The Death of the Sensuous Chemist: The 'New' Chemistry and the Transformation of Sensuous Technology." In *Empire of the Senses: The Sensual Culture Reader,* edited by David Howes, 503–529. Oxford: Berg, 2005.

Roberts, Wm. "Appendix D: Distant Control of Valves." *Proceedings of the American Electric Railway Engineering Association* 11 (1913): 453–461.

Roedding, Gordon E. "Push Button." Patent No. 1,135,926. April 13, 1915.

Rollins, Judith. *Between Women: Domestics and Their Employers.* Philadelphia: Temple University Press, 1985.

Rowand, Lewis G. "Fire-Alarm System." Patent No. 477,068. 1892.

Royal Easy Chairs. "Push the Button—and Rest." *McClure's Magazine* 44 (1914): 142.

Rubinow, I. M., and Daniel Durant. "The Depth and Breadth of the Servant Problem." *McClure's Magazine* 34, no. 5 (1910): 576–585.

Russell, Frances Eldredge. "Foundation and Fellowship." *Arena* 15 (1896): 960–964.

Ryan, Katy. *Demands of the Dead: Executions, Storytelling, and Activism in the United States*. Iowa City: University of Iowa Press, 2012.

"A Safety Device." *Architect and Engineer* 13 (1908): 84.

Salisbury, Albert. *The Theory of Teaching and Elementary Psychology*. Whitewater, WI: The Century Book Company, 1905.

Salomons, David. *Electric Light Installations & the Management of Accumulators*. London: Whitaker & Co., 1890.

Sarukkai, Sundar. "Phenomenology of Untouchability." *Economic and Political Weekly* 44, no. 37 (2009): 39–48.

Schaeffer, N. C. "Grand Opening Ceremony." *Pennsylvania School Journal* 42 (1893): 13.

Schapiro, J. Salwyn. *Modern Contemporary European History*. Boston: Houghton Mifflin Company, 1918.

Scharff, Virginia. *Taking the Wheel: Women and the Coming of the Motor Age*. Toronto, ON: Free Press, 1991.

Schauffer, Adolphus Frederick. *Select Notes on the International Sunday School Lessons*. Cambridge, MA: W. A. Wilde Company, 1911.

Schivelbusch, Wolfgang. *Disenchanted Night: The Industrialization of Light in the Nineteenth Century*. Berkeley: University of California Press, 1988.

Schivelbusch, Wolfgang, and Damian J. Kulash. *The Railway Journey: Trains and Travel in the 19th Century*. New York: Urizen Books, 1979.

Schlereth, Thomas J. *Victorian America: Transformations in Everyday Life, 1876–1915*. New York: HarperCollins, 1991.

Schnecker, Joseph E. "In the Chain of Bureaucracy." *Popular Electricity and the World's Advance* 3 (1910): 563.

Schneider, George A. "Technical Hints." *Journal of Electricity* 39, no. 11 (1917): 517–518.

Schuylkill Electric Const. & Supply. "Specifications for Electrical Work for Mr. Simon Krick." 1909. Warshaw Collection of Business Americana. Archives Center, National Museum of American History, Smithsonian Institution, Washington, DC.

Sconce, Jeffrey. *Haunted Media: Electronic Presence from Telegraphy to Television*. Durham, NC: Duke University Press, 2000.

Scott, Roscoe Gilmore. "Let Us Go Back." *Edison Monthly* 6, no. 4 (1913): 148. Warshaw Collection of Business Americana. Archives Center, National Museum of American History, Smithsonian Institution, Washington, DC.

Scrutton, Percy E. *Electricity in Town and Country Houses*. London: Archibald Constable and Co., 1898.

Sears, Roebuck and Co. *Electrical Goods and Supplies*. Chicago: Sears, Roebuck and Co., ca. 1902. Collections of the Bakken Museum, Minneapolis, MN.

Seelman, M. S., Jr. "Electric Light Jingles by Edison Electric Illuminating Co. of Brooklyn." 1908. Warshaw Collection of Business Americana. Archives Center, National Museum of American History, Smithsonian Institution, Washington, DC.

Segrave, Kerry. *Vending Machines: An American Social History*. Jefferson, NC: McFarland, 2002.

Semi-Weekly Tribune. "May He Touch the Right Button." May 11, 1897: 4.

Sentinel. "A Practical Joker." May 15, 1887: 4.

Setoodeh, Ramin, and Elizabeth Wagmeister. "Matt Lauer Accused of Harassment by Multiple Women." *Variety*, November 29, 2017.

http://variety.com/2017/biz/news/matt-lauer-accused-sexual-harassment-multiple-women-1202625959/.

Severy, Hazel W. "Applied Science as the Basis of the Girl's Education." *Journal of Proceedings and Addresses of the National Education Association of the United States* 53 (1915): 1020–1021.

Sewell, William H., Jr. "Toward a Post-Materialist Rhetoric for Labor History." In *Labor History: Essays on Discourse and Class Analysis*, edited by Lenard R. Berlanstein, 15–38. Champaign-Urbana: University of Illinois Press, 1993.

Shepardson, Geo. D. *Electrical Catechism: An Introductory Treatise on Electricity and Its Uses*. New York: American Electrician Co., 1901.

Shermer, Michael. "What Milgram's Shock Experiments Really Mean." http://www.scientificamerican.com/article.cfm?id=what-milgrams-shock-experiments-really-mean/.

Sherwood, Mary Elizabeth Wilson. *Amenities of Home*. New York: D. Appleton and Company, 1881.

Simon, Linda. *Dark Light: Electricity and Anxiety from the Telegraph to the X-Ray*. Orlando, FL: Harcourt, 2004.

Sleeper, M. B. *Electric Bells: A Handbook to Guide the Practical Worker in Installing, Operating, and Testing Bell Circuits, Burglar Alarms, Thermostats and Other Apparatus Used with Electric Bells*. New York: Norman W. Henley, 1917.

Sloane, Thomas O'Conor. *Electric Toy Making for Amateurs*. New York: Norman W. Henley, 1915.

Smith, Nelson. "A Partial History of Alarms." *Baffler* 10 (1997): 25–30.

Sohon, Michael D. "Chemistry in Secondary Schools." *Science* 31, no. 808 (1910): 979–983.

Solis-Cohen, Solomon. *A System of Physiologic Therapeutics: Prophylaxis, Personal Hygiene, Civic Hygiene, and Care of the Sick*. Vol. 5. Philadelphia: P. Blakiston's Son & Co., 1903.

Sopee, George J. "Electric Push-Button." Patent No. 559,416. Filed May 17, 1895; issued May 5, 1896.

Sorokanich, Bob. "Bosch Is Working on Touchscreen Buttons You Can Actually Feel." *Road & Track*, November 11, 2015. http://www.roadandtrack.com/new-cars/car-technology/news/a27320/bosch-is-working-on-touchscreen-buttons-you-can-actually-feel. Accessed May 1, 2016.

Sparks, William. "Electric Push-Button." Patent No. 1,239,054. Filed April 25, 1913; issued September 4, 1917.

Spon, Edward, and Francis N. Spon. *Spons' Mechanics' Own Book: A Manual for Handicraftsmen and Amateurs.* London: E. & F. N. Spon, 1886.

Sprague, Frank J. "Electric Elevators with Detailed Description of Special Types." Paper presented at the 102nd meeting of the American Institute of Electrical Engineers. 1896. Warshaw Collection of Business Americana. Archives Center, National Museum of American History, Smithsonian Institution, Washington, DC.

Srinivasan, Karthik. "The Aftermath of the Buttonization of Our Emotions." *social@Ogilvy*, March 1, 2016. https://social.ogilvy.com/the-aftermath-of-the-buttonization-of-our-emotions.

Stahl, William A. "Venerating the Black Box: Magic in Media Discourse on Technology." *Science, Technology & Human Values* 20, no. 2 (1995): 234–258.

Stallard, Mrs. Arthur. *The House as Home.* New York: James Pott & Company, 1913. Collections of the Winterthur Library, Winterthur, DE.

Standage, Tom. *The Victorian Internet: The Remarkable Story of the Telegraph and the Nineteenth Century's On-Line Pioneers.* New York: Walker and Co., 1998.

Stanley, Elijah H. "Electric Push-Button." Patent No. 591,895. 1897.

Star, S. L., and K. Ruhleder. "Infrastructures Are Also Relational: Steps toward an Ecology of Infrastructure: Design and Access for Large Information Spaces." *Information Systems Research* 7, no. 1 (1996): 111–134.

Star, Susan Leigh, and Anselm Strauss. "Layers of Silence, Arenas of Voice: The Ecology of Visible and Invisible Work." *Computer Supported Cooperative Work* 8 (1999): 9–30.

Steele, G. F. *Electricity in a Modern Residence*. New York: H. Ward Leonard & Co., 1892.

Steele, James W. *Steam, Steel and Electricity*. Chicago: The Werner Company, 1895.

Steinmetz, Charles P. "Back of the Electric Button." *Good Housekeeping* 76, no. 5 (1923): 48.

Steinmetz, Charles P. "The A-B-C's of Electricity." *Popular Science* 101, no. 1 (1922): 29–30, 109–110.

Sterne, Jonathan. *The Audible Past: Cultural Origins of Sound Reproduction*. Durham, NC: Duke University Press, 2003.

Stewart, George Walter. "A Contribution of Modern Physics to Religious Thought." *Homiletic Review* 68 (1914): 273–279.

Stich, Herman J. "Push-Button Gents." *Los Angeles Times*, January 11, 1923: II4.

Still, Andrew Taylor. *Osteopathy, Research and Practice*. Kirksville, MO: A. T. Still, 1910.

St. John, M. Thomas. *Things a Boy Should Know about Electricity*. New York: Hard Press, 1900.

Stone, L. A., ed. *Sex Searchlights and Sane Sex Ethics: An Anthology of Sex Knowledge*. Chicago: Science Publishing Company, 1922.

Stott, Richard Briggs. *Jolly Fellows: Male Milieus in Nineteenth-Century America*. Baltimore: Johns Hopkins University Press, 2009.

The Stout-Meadowcroft Company. *Illustrated Catalogue and Price List of the Stout-Meadowcroft Co.* New York: Stout-Meadowcroft, 1885. Collections of the Bakken Museum, Minneapolis, MN.

Strasser, Susan. *Never Done: A History of American Housework*. New York: Pantheon Books, 1982.

Straus, Max. "Electric Push-Button." Patent No. 452,397. May 19, 1891.

Stromberg-Carlson Telephone Company. "Don't Walk—Push *One* Button *Once*—and *Talk*." *Textile World* 60 (1922): 1336.

Strong, Anna Louise. *Child-Welfare Exhibits: Types and Preparation*. Washington, DC: Government Printing Office, 1915.

Sturken, Marita, Douglas Thomas, and Sandra Ball-Rokeach. *Technological Visions: The Hopes and Fears That Shape New Technologies*. Philadelphia: Temple University Press, 2004.

Sullivan, J. F. "The End of War." *Strand Magazine* 3 (1892): 644–649.

Sully, James. "Pushing." In *Good Words for 1880*, edited by Donald Macleod. London: Isbister and Company, 1880.

Sun. "Electric Fire Alarm; Fifty Years since the Inauguration of the System." May 1, 1902.

Sun. "The Major Sat on the Push-Buttons." April 11, 1904: 11.

Sun. "New Fire-Alarm System: It Is Designed to Meet Needs of Small Cities." September 17, 1906: 10.

Sun. "Once Common, Now Rare." October 17, 1909: 17.

Sun. "Push-Button Inventor Dead." August 17, 1907: 5.

Sun. "200 Captured in Raid." August 28, 1905: 5.

Sun. "Winter Cars Running." October 15, 1898: 12.

Sydell, Laura. "Microsoft's Kinect Brings Gesture to a New Level." *NPR*, November 4, 2010. http://www.npr.org/templates/story/story.php?storyId=131074438.

Tate, Mrs. C. B., Jr. "The Field of Women in Public Relations." *National Electric Light Association Bulletin* 9 (1922): 249–250.

Taylor, Frederick Winslow. *The Principles of Scientific Management*. New York: Harper & Brothers, 1913.

Thompson, Emily Ann. *The Soundscape of Modernity: Architectural Acoustics and the Culture of Listening in America, 1900–1933*. Cambridge, MA: MIT Press, 2002.

Timmons, Todd. *Science and Technology in Nineteenth-Century America*. Westport, CT: Greenwood Press, 2005.

Toller, Rev. T. N. *Sermons on Various Subjects*. London: B. J. Holdsworth, 1824.

Town and Country. "Stow-Away Places in Touring Cars: Novel Devices for Packing Impedimenta on Long Trips." November 3, 1906: 24. Warshaw Collection of Business Americana. Archives Center, National Museum of American History, Smithsonian Institution, Washington, DC.

Town and Country. "What Has Become of the Door Plates?" May 19, 1906: 29.

Traynor, Des. "On Magical Software." *Inside Intercom* (blog), July 2, 2014. http://blog.intercom.io/on-magical-software/.

Trevert, Edward. *How to Make and Use an Electric Bell*. Lynn, MA: Bubier Publishing Co., 1906.

"Tric Gear Shift." *Automobile Topics* 34 (1914): 223.

Trump, Donald J. (@realDonaldTrump). "North Korean Leader Kim Jong Un just stated that the 'Nuclear Button is on his desk at all times.' Will someone from his depleted and food starved regime please inform him that I too have a Nuclear Button, but it is a much bigger & more powerful one than his, and my Button works!" Twitter, January 2, 2018, 4:49 p.m. https://twitter.com/realdonaldtrump/status/948355557022420992.

Tunzelmann, G. W. *Electricity in Modern Life*. New York: P. F. Collier & Son, 1902.

Turner, Bryan S. *The Body and Society: Explorations in Social Theory*. Thousand Oaks, CA: Sage, 2008.

"Untitled." *Electrical World* 31 (1898): 266.

"Untitled." *Electrician* 41, no. 22 (1898): 703.

"Untitled." *Hardware Dealers' Magazine* 5 (1896): 240.

Urquhart, John W. *Electric Light Fitting.* London: Crosby Lockwood and Son, 1890.

Vail, Charles Henry. *Principles of Scientific Socialism.* Chicago: Charles H. Kerr & Company, 1899.

Van Amburgh, Fred DeWitt. *The Buck Up Book.* New York: The Silent Partner Co., 1919.

Vanhemert, Kyle. "Apple's Haptic Tech Is a Glimpse at the UI of the Future." *Wired*, March 19, 2015. http://www.wired.com/2015/03/apples -haptic-tech-makes-way-tomorrows-touchable-uis/. Accessed January 15, 2016.

"Vertical Transportation." *Scientific American* 88, no. 14 (1903): 238.

"The Victory of the Push Button." *Appeal* 24, no. 13 (1908): n.p.

Vidyarthi, Guru Datta. *Works: With a Biographical Sketch.* Lahore, Pakistan: Aryan Printing, Publishing & G. Trading Co., 1902.

"The Vulcan Electric Gear Shift." *Motor World Wholesale* 38 (1913): 59.

Wagstaff, Keith. "Project Glass: Google's Augmented Reality Glasses Take Hands-Free Computing to the Extreme." *Time*, April 4, 2012. http://techland.time.com/2012/04/04/googles-augmented-reality -glasses-take-hands-free-computing-to-the-extreme/.

Walker, Rob. "Ad Play." *New York Times Magazine*, December 17, 2006. http://www.nytimes.com/2006/12/17/magazine/17wwln_consumed .t.html.

Wall, James W. "Address of James W. Wall." In *Appendix to the Journal of the Sixteenth Senate of the State of New Jersey*, 555–580. Belvidere, NJ: John Simerson, 1860.

Wallace, A. R. "Immortality and Morality." *Borderland* 2 (1895): 10.

Walmsley, R. Mullineux. *The Electric Current: How Produced and How Used.* London: Cassell and Company, Limited, 1894.

Ward, Leonard H. *Electricity in a Modern Residence*. New York: H. Ward Leonard & Co., 1892.

Warder, George Woodward. *The Cities of the Sun*. New York: G. W. Dillingham, 1901.

Warshaw, Jacob. *The New Latin America*. New York: Thomas Y. Crowell Company, 1922.

Washington Post. "An Automatic Milkman." May 3, 1908.

Washington Post. "The Bored Woman's Lament." February 18, 1906: P4.

Washington Post. "A City 100 Years from Now: The Luxuries People of the Next Century Will Probably Enjoy." March 5, 1911: MS4.

Washington Post. "Elevators for a Palace." September 8, 1901: 22.

Washington Post. "Fingers Cut Off by Wheels." July 9, 1902: 2.

Washington Post. "Home a Hundred Years Hence: Some of the Wonderful Improvements People of the Next Century Will Enjoy." October 27, 1907: M2.

Washington Post. "The Honeymoon Car." February 3, 1906: 3.

Washington Post. "The Informator." October 18, 1933: 6.

Washington Post. "Masonic Fair Is Open: President Roosevelt Set Convention Hall Aglow." April 15, 1902: 11.

Washington Post. "New Page Call in Operation." December 25, 1901: 9.

Washington Post. "A Nickel in the Slot." April 19, 1891.

Washington Post. "Opened the Fair." June 2, 1905: 1.

Washington Post. "Push-Buttons on Street Cars." April 8, 1894: 10.

Washington Post. "Spot Pressed the Button." April 1, 1894: 16.

Washington Post. "Wants to Press the Button." February 9, 1892: 1.

Watrous. "Push the Button and We Do the Rest." *Domestic Engineering and the Journal of Mechanical Contracting* 51, no. 13 (1910): 9.

Waugh, F. A. "Landscape Photography." *Photo-Miniature* 3, no. 1 (1902): 1–36.

Weaver, Bent L., and William R. Miller. "Push-Button System for Desks and the Like." Patent No. 1,003,677. September 19, 1911.

Webb, Herbert Laws. "The $100 Prize Essay. How Can the Department of Electricity of the World's Columbian Exposition Best Serve the Electrical Interests?" *World's Fair Electrical Engineering* 1 (1893): 59–63. Warshaw Collection of Business Americana. Archives Center, National Museum of American History, Smithsonian Institution, Washington, DC.

Webb, Herbert Laws. "The Future of the Telephone Industry." *Engineering Magazine* (1892): 753–761.

Weber, E. D. *Practical Wiring for Buildings: For Incandescent Electric Lighting, Electric Gas Lighting, Electric Burglar Alarms, Electric House and Hotel Annunciators, Bells, Etc., Etc.* New York: The Comenius Press, 1895.

Wedgwood, Hensleigh. *Dictionary of English Etymology.* New York: Macmillan, 1878.

Weekly Register-Call. "To Call a Policeman." July 19, 1889.

Westall, Carroll. *The House Electrical; Being a Brief Description of the Ideal Home and How to Plan and Equip It.* Boston: Pettingell-Andrews, 1912. Collections of the Winterthur Library, Winterthur, DE.

Western Electric Company. "Make Your Push Button an Office Telephone." *Literary Digest* 49, no. 16 (1914): 756.

Western Electric Company. "Intercommunicating Telephone Systems." 1908. Warshaw Collection of Business Americana. Archives Center, National Museum of American History, Smithsonian Institution, Washington, DC.

Western Electric Manufacturing Co. *Electric Bells, Annunciators, Burglar Alarms, Electro-Mercurial Fire Alarm and Electric Gas Lighting.* 1882. Collections of the Huntington Library, San Marino, CA.

Westinghouse Electric & Manufacturing Co. "Your Unseen Servant." *Popular Mechanics* 28, no. 1 (1917): 102.

Whipple, Elizabeth. "Some Electrical Conveniences Women Want Contractors to Provide." *National Builder* 62 (1919): 52.

Whitman, Roger B. "Planning for the Wiring of the House." *Country Life* 40 (1921): 62–63.

Willard Storage Battery Co. "Untitled." *Motor Age* 24 (1911): 49.

Williams, Archibald. *How It Works; Dealing in Simple Language with Steam, Electricity, Light, Heat, Sound, Hydraulics, Optics, Etc. And with Their Application to Apparatus in Common Use.* New York: Thomas Nelson & Sons, ca. 1910.

Williams, Henry Smith, and Edward H. Williams. *Every-Day Science.* Vol. 9. New York: The Goodhue Company, 1910.

Williams, Henry Smith, and Edward Huntington Williams. *A History of Science: The Conquest of Nature.* Vol. 9. New York: Harper & Brothers, 1910.

Williams, Henry Smith, and Edward Huntington Williams. "The Modern Sky-Scraper." *Wonders of Science in Modern Life* 5 (1912): 123.

Wines, Arthur F. "Coin Controlled Machinery." *Sibley Journal of Engineering* 13 (1899): 219–220.

Winner, Langdon. *Autonomous Technology: Technics-out-of-Control as a Theme in Political Thought.* Cambridge, MA: MIT Press, 1977.

Winterburn, George William. "Microbes and Disease. Simple Explanations." *Everybody's Magazine* 3 (1900): 449–457.

Wittbecker, William A. "Domestic Electrical Work." *Sanitary and Heating Age (The Metal Worker)* 43, no. 14 (1895): 86.

Woodcroft, Bennet. *Chronological and Descriptive Index of Patents Applied for and Patents Granted.* London: George Edward Eyre and William Spottiswoode, 1874.

‍

Woolnough, Jeff, dir. *The Outer Limits.* Season 3, episode 13, "Dead Man's Switch." Aired April 4, 1997, on Showtime.

Wright, John. *The Home Mechanic: A Manual for Industrial Schools and Amateurs.* New York: E. P. Dutton and Company, 1905.

Wright, John. "Bell-Hanging for Inside Rooms." *Building Age* 6, no. 2 (1884): 28.

Wright, Megh. "Seth Meyers Takes 'A Closer Look' at the Matt Lauer Sexual Misconduct Allegations." *SplitSider*, December 1, 2017. http://splitsider.com/2017/12/seth-meyers-takes-a-closer-look-at-the-matt-lauer-sexual-misconduct-allegations/.

Wu, Tim. "The Problem with Easy Technology." *New Yorker*, February 21, 2014. http://www.newyorker.com/tech/elements/the-problem-with-easy-technology.

Wyman, Ferdinand A. "Constitutionality of Execution by Electricity." In *Proceedings of the National Electric Light Association*, 125–159. New York: James Kempster Printing Company, 1890.

Yates, Joanne. *Control through Communication: The Rise of System in American Management.* Baltimore: Johns Hopkins University Press, 1989.

Young, H. W. "Remote-Control Switches." *Electrical Review* 53, no. 21 (1908): 790–791.

Young, Henry Walter. "Electric Automobile Horn." *Popular Electricity and the World's Advocate* 2 (1909): 746.

"'Yours to Command.' New Movie Film of N.E.L.A." *Electric Traction* 18 (1922): 636.

Zandy, Janet. *Hands: Physical Labor, Class, and Cultural Work.* New Brunswick, NJ: Rutgers University Press, 2004.

Zenith Radio Corporation. "Why You Can Operate Zenith TV from Your Easy Chair." *Coronet*, February 1951. Retrieved from Duke Ad*Access.

Zuboff, Shoshana. *In the Age of the Smart Machine: The Future of Work and Power.* New York: Basic Books, 1988.

Index

Printed in the United States
by Baker & Taylor Publisher Services